Annals of Mathematics Studies

Number 132

Commensurabilities among Lattices in $PU(1,n)$

by

Pierre Deligne and G. Daniel Mostow

PRINCETON UNIVERSITY PRESS

———

PRINCETON, NEW JERSEY

1993

Princeton University Press books are printed on acid-free paper
and meet the guidelines for permanence and durability of the Committee on Production
Guidelines for Book Longevity of the Council on Library Resources

Printed in the United States of America

Library of Congress Cataloging-in-Publication Data

Deligne, Pierre.
      Commensurabilities among lattices in $PU(1,n)$ / by Pierre
Deligne and G. Daniel Mostow.
            p.   cm. — (Annals of mathematics studies ; no. 132)
      Includes bibliographical references.
      ISBN 0-691-03385-4 — ISBN 0-691-00096-4 (pbk.)
      1. Functions, Hypergeometric. 2. Monodromy groups.
      3. Lattice theory. I. Mostow, George D.   II. Title.
      III. Series
QA353.H9D45   1993
515'.25—dc20   93-5528

The publisher would like to acknowledge the authors of this volume for providing the
camera-ready copy from which this book was printed

CONTENTS

ACKNOWLEDGMENTS

The authors are grateful to Dr. Rolf-Peter Holzapfel whose careful reading of a preliminary version of this monograph helped us eliminate a number of misprints. The authors wish to acknowledge our debt to Ms. Donna Belli whose expertise and patience transformed our manuscript and diagrams into camera-ready form.

Commensurabilities among Lattices in $PU(1,n)$

§1. INTRODUCTION

The aim of this monograph is to investigate lattices in $PU(1,n)$ groups.

Fix $n \geq 1$ and let Q be the moduli space of $(n+3)$-uples of distinct points on a projective line: the quotient by $PGL(2)$ of $(\mathbf{P}^1)^{n+3}$ minus the diagonals $x_i = x_j$. Rather than dividing by $PGL(2)$, one may fix $x_{n+1} = 0, x_{n+2} = 1, x_{n+3} = \infty$. This identifies Q with the space M of n-uples $(x_1, \ldots, x_n)$ of distinct points of $\mathbf{P}^1$, with $x_i \neq 0, 1, \infty$.

Let μ be a $(n+3)$-uple of complex numbers with sum 2. In our 1986 paper, referred to hereafter as [DM], we defined the vector space $V(\mu)$ of integrals over suitable cycles of

$$w_\mu = \prod_{1}^{n+2} (z - x_i)^{-\mu_i} dz,$$

where the integrals are viewed as functions of x (with $x_{n+1} = 0, x_{n+2} = 1, x_{n+3} = \infty$). It is a $(n+1)$-dimensional vector space of multivalued functions on Q, identified with M. In [Lauricella 1893], $V(\mu)$ is described as the space of solutions of a system of linear partial differential equations expressing all second derivatives in terms of derivatives of order ≤ 1.

For g a fixed multivalued function on M of the form

$$\Pi(x_i - x_j)^{a_{ij}} \qquad (1 \leq i \leq n, 1 \leq j \leq n+2, i < j)$$

the multivalued functions $f \cdot g$ for $f \in V(\mu)$ form again a vector space W of dimension $(n+1)$. Such a twist of a $V(\mu)$ we call *hypergeometric-like* . The consideration of all twists of $V(\mu)$ restores the symmetry among the indices $1, \ldots, n+3$ which was broken by imposing $x_{n+1} = 0, x_{n+2} = 1, x_{n+3} = \infty$: any permutation $\sigma \in \Sigma(n+3)$ acts on Q, and the transform of $V(\mu)$ by σ is a twist of $V(\sigma(\mu))$.

Any $V(\mu)$, or one of its twists, is the space of global sections on a universal covering of Q, of a local system of holomorphic functions on Q; for $V(\mu)$: the local system of solutions of Lauricella's equations. The local systems corresponding to twists of $V(\mu)$ we call as before hypergeometric-like.

Let Q^+ be the partial compactification of Q obtained by allowing at most two of the $n+3$ points x_i to coalesce. The complement of Q in Q^+ is a union of disjoint irreducible divisors $D_{s,t}, s \neq t$.

The sections 2 to 7 are devoted to a characterization of hypergeometric-like local systems on Q by properties which are local on Q^+. Our results are related to those of Picard and Terada. For a description of Terada's results see §7.13.

For W a twist of $V(\mu)$, the monodromy group Γ' is a subgroup of $GL(W)$. Let Γ be the projective monodromy group: the image of Γ' in $PGL(W)$. The projective monodromy group does not change by twisting, hence depends only on μ and is denoted Γ_μ. When all the μ's are real between 0 and $1(0 < \mu_i < 1)$, the monodromy preserves an hermitian form of signature $(1,n)$ on $V(\mu)$ and Γ_μ is a subgroup of $PU(1,n)$, well defined up to conjugacy.

If μ is invariant by a subgroup H of the symmetric group $\Sigma(n+3)$, suitable twists W of $V(\mu)$ will be invariant by H. The local system of functions W then descends as a local system of functions W/H on Q/H - or rather on Q'/H, where Q' is the open subset of Q where the action of the quotient $\bar{H}$ of H acting effectively on Q is free. For $n = 1, \bar{H}$ is the image of H in $\Sigma(4)/$ Vierergruppe. For $n > 1, \bar{H} = H$.

The projective monodromy group $\Gamma_{\mu,H}$ of W/H depends only on μ and H; it is an extension of a quotient of H by Γ_μ.

For $n = 2$, we construct birational maps $\varphi : Q \to Q/H_2$, for suitable subgroups H_2 of $\Sigma(5)$ and apply the characterization of §7 to show that the pull back of suitable hypergeometric-like local systems W/H_2 on Q/H_2 are again hypergeometric-like. For W a twist of $V(_2\mu)$ and the pull back a twist of $V(_1\mu)$, this identifies $\Gamma_{_1\mu}$ with a subgroup of $\Gamma_{_2\mu H_2}$, of index dividing the degree of φ. If φ factors through Q/H_1, we similarly get $\Gamma_{_1\mu H_1} \hookrightarrow \Gamma_{_2\mu H_2}$.

Examples of the commensurabilities we treat are (cf. (10.6.1)):

[A]   Let $_1\mu = (a, a, b, b, 2-2a-2b)$, $_2\mu = (1-b, 1-a, a+b-\frac{1}{2}, 1-a-b)$

$$H_1 = \langle(1,2),(3,4)\rangle, H_2 = \langle(3,4)\rangle$$

[B]   Let $\frac{1}{\rho} + \frac{1}{\sigma} = \frac{1}{6}$,

$$_1\mu = (\frac{1}{2} - \frac{1}{\rho}, \frac{1}{2} - \frac{1}{\rho}, \frac{1}{2} - \frac{1}{\rho}, \frac{1}{6} + \frac{1}{\rho},$$

$$2(\frac{1}{6} + \frac{1}{\rho})), \quad _2\mu = (\frac{1}{6}, \frac{1}{6}, \frac{1}{6}, \frac{5}{6} - \frac{1}{\rho}, \frac{5}{6} - \frac{1}{\sigma}).$$

In situations as above, $\Gamma_{1\mu}$ will be a lattice in $PU(1, n)$ if and only if $\Gamma_{2\mu}$ is. The projective monodromy group Γ_μ $(0 < \mu < 1)$ is not dense in $PU(1, 2)$ only when it is a lattice, and then by the results of Mostow [M4] §9, 1988 and the present paper, either

(a) Γ_μ is arithmetic, or

(b) μ satisfies condition Σ INT of [M 3], or

(c) Γ_μ is related to another Γ_μ, satisfying (b), by the previous construction. We were in fact led to the construction of the relevant maps $\varphi : Q \to Q/H$ by trying to understand some commensurability results between Γ_μ's obtained by J.K. Sauter in his 1988 dissertation after some computer exploration. Those results were needed in order to decide which of the Γ_μ's were discrete.

In the groups $O(1, n)$, there are, for each n, infinitely many commensurability classes of non arithmetic groups (Gromov-Piatetski-Shapiro [GP] 1988). In the other simple groups, except those of the series $SU(1, n)$, they are none (Corlette, Gromov-Schoen). In $SU(1, n)$, non arithmetic lattices have been found so far only for $n \leq 3$, with only one example (up to commensurability) for $n = 3$. The following approaches have been used to look for possibly non arithmetic lattices:

I. Take the subgroup of $U(1, n)$ generated by suitable complex reflections

II. Take a $\Gamma(\mu)$, monodromy group of some Lauricella hypergeometric function.

III. Take a familiar projective algebraic variety - for instance $\mathbf{P}^2$ - a configuration of hypersurfaces on it and assign weights to them, to define an orbifold. Apply Yau's theorem or variants thereof to the orbifold to claim it is the quotient of the complex ball by a lattice Γ in $PU(1, n)$.

Surprisingly many of the lattices constructed so far by these methods are commensurable to a Γ_μ. The type I lattices constructed in [M2] are of type II by the results in [M3]. The type II lattice, when the condition INT of [DM], or more generally Σ INT of [M 3] is satisfied, are special cases of type III, related to the complete quadrangle in $\mathbf{P}^2$, and variants thereof: blowing up and down, and quotients by finite groups. In §§15, 16 we consider a number of type III construction, for surfaces, and show how they can be reduced to type II lattices. If R is the relevant orbifold, the method each time is to replace R by a quotient R/H_2, and to construct a map $Q/H_1 \to R/H_2$ ($H_1 \subset \Sigma(5)$), extending for a suitable H-invariant μ, to $\overline{\varphi} : Q_\mu^{st}/H_1 \to R/H_2$. As μ satisfies ΣINT, Q_μ^{st}/H_1 is a quotient of the complex ball: $B/\Gamma_{\mu H_1}$. The map $\overline{\varphi}$ is a covering map, in the orbifold sense, so that if R is a quotient of the ball by Γ, R/H_2 is a quotient by a bigger group Γ', and $\Gamma_{\mu H_1}$ is of finite index in Γ'.

Both here, and for isogenies between Γ_μ, one source of the required maps φ is the theory of Lauricella hypergeometric functions in the elliptic and parabolic cases ([DM]).

The characterization Theorem 7.1 for hypergeometric like functions can be paraphrased as follows.

Let V be a local system of holomorphic functions on Q with $dim\ V = 1 + dim\ Q$. Assume that

(a) V is "non-degenerate" in the sense that at each point of Q, a base of V provides an etale map of a neighborhood to $\mathbf{P}^n$, and that

(b) at each divisor $D_{s,t}$ for $\{s,t\} \in \binom{S}{2}$, the system V has strict exponents $(\alpha_{s,t}, \beta_{s,t})$ in the sense explained in §6.9. Then (cf. Proposition 7.5)

(i) $\beta_{s,t} = \sum\limits_{\{u,v\}} \alpha_{u,v}$ (sum over $(\binom{S - \{s,t\}}{2})$)

(ii) $\sum\limits_{\{s,t\}} \alpha_{s,t} = 1$ (sum over $(\binom{S}{2})$)

and

(iii) V is the twist of a hypergeometric local system on Q.

This theorem is related to Riemann's classical theorem:

A multivalued function on the projective line punctured at c_1, c_2, c_3:

$$
F \begin{pmatrix} c_1 & c_2 & c_3 & \\ \alpha_1 & \alpha_2 & \alpha_3 & x \\ \beta_1 & \beta_2 & \beta_3 & \end{pmatrix}
$$

with regular singularities of exponent (α_i, β_i) at $c_i, i \in \mathbf{Z}/3\mathbf{Z} = \{i = 1, 2, 3\}$

with any three of its branches linearly dependent, and with $\sum_{i=1}^{3}(\alpha_i+\beta_i) = 1$, satisfies the second order equation

$$y'' + y' \sum_i \frac{1 - \alpha_i - \beta_i}{x - c_i} + \frac{y}{\prod_i(x - c_i)} \sum_i \frac{\alpha_i \beta_i (c_i - c_{i-1})(c_i - c_{i+1})}{(x - c_i)}.$$

If $(c_1, c_2, c_3) = (0, 1, \infty)$ and $\alpha_0 = \alpha_1 = 0$, we have the classical hypergeometric equation. From our view point, the hypothesis $\sum_i (\alpha_i + \beta_i) = 1$ is equivalent to our non-degeneracy condition on Q^+, which in this case, $n = 1$, is a non-Hausdorff manifold! This hypothesis is precisely condition (ii) above.

The proof of Theorem 7.1 entails first proving (i), (ii) and thereafter (iii). Once (i) and (ii) are established, we can by 3.5 twist the given local system to obtain a new local system whose exponents match those of the local system of Lauricella's hypergeometric functions. We then could appeal to a uniqueness theorem of Terada [1973] stating that the Lauricella hypergeometric functions are uniquely determined by their exponents $(\alpha_{s,t}, \beta_{s,t})\{s, t\} \in \binom{S}{2} - \{\{a, b\}, \{b, c\}, \{c, a\}\}$ where $(x_a, x_b, x_c) = (0, 1, \infty)$. In the interest of a self-contained account, we have taken another route borrowing from the theory of hypergeometric functions on Grassmannian spaces developed by Gelfand and his collaborators.

By comparing exponents, Theorem 7.15 can be used to obtain identities among hypergeometric functions of n variables in much the same way that for $n = 1$, the above cited Riemann's theorem can be used to prove the identities that Kummer had proved by formal methods 20 years before Riemann's Theorem. We give an explicit example of such an identity in (13.6.2). The resulting identity between integrals can be reformulated as an identity in a power series expansion of functions of two variables. Upon matching coefficients, one can then obtain a classical identity (Pfaff, 1797) expressing $_3F_2\left(\begin{array}{ccc} -n & a & b \\ & c & d; \; 1 \end{array}\right)$; retracing steps would then yield (13.6.2).

The paper has been arranged so that results on exponents of hypergeometric functions can be read in the first part §§2-7.

§2 presents results we require about divisors in an algebraic variety over the field of complex numbers $\mathbf{C}$.

§3 deals with results needed for the existence of multivalued functions on Q with prescribed valuations at the divisors $D_{s,t}, \{s,t\} \in \binom{S}{2}$. It also gives information about $Pic\ Q^+$ needed in the discussion of exponents.

§4 defines local systems of holomorphic functions and the Lauricella local system hypergeometric functions in particular.

Twists of Lauricella local systems are called hypergeometric-like. The hypergeometric-like local systems on Q are related to Gelfand's hypergeometric local systems defined on a Zariski-open set of the space $\bar{H}$ of $N \times 2$ matrices. Indeed our discussion of the relation of $Pic\ Q^+$ to $Pic\ H^+$, where H^+ is the open subset of $\bar{H}$ corresponding to Q^+, begins already in §3.12.

§5 describes the relation of hypergeometric-like local systems on Q to Gelfand local systems on H and derives in Proposition 5.11, for any f in a hypergeometric-like local system, a set of linear differential equations expressing all second derivatives of f in terms of f and its first derivatives.

§6 makes precise the notion of non-degeneracy of exponents for a $(1 + dim\ Q)$-dimensional local system of holomorphic functions on Q. The basic notion is that of *strict exponents* .

§7 begins with Theorem 7.1 which asserts that any "non-degenerate" local system of Q possessing strict exponents at $D_{s,t}$ for all $\{s,t\} \in \binom{S}{2}$ is necessarily hypergeometric-like. The proof occupies most of the section (7.1-7.13). The remaining sections explain our remarks above about Riemann's case $n = 1$, and organizes the results on hypergeometric-like functions in a form needed for the commensurability applications.

§8 recapitulates preliminaries drawn from [DM] and [M 3] about the groups $\Gamma_{\mu H}$ and adds some information about fixed points of the action of elements of $\Sigma(S)$ on Q.

§9 explains how the problem of commensurability of $\Gamma_{i\mu H_i}$ $(i = 1, 2)$ arose, sketches the results in J.K. Sauter's thesis based on the presentation of $\Gamma_{\mu\Sigma(3)}$ given in [M 2], and presents our strategy for reproving his results via geometric methods.

§10 applies the strategy afforded by §7 to two cases via isomorphisms

$$_1\bar{Q}/H_2 \longrightarrow_2 \bar{Q}/H_2$$

where $_1\bar{Q}$ and $_2\bar{Q}$ are suitable compactifications of Q. In §11, the case [B] above is treated. In case ρ, σ are integers, obtaining $\Gamma_{_1\mu\Sigma(3)}$ as a subgroup of $\Gamma_{2\mu\Sigma(3)}$ whether by Sauter's method, or by ours is the only way so far to prove that $\Gamma_{1\mu\Sigma(3)}$ is discrete in $PU(1,2)$ (and hence a lattice by [DM;§§11]). The finite map

$$\varphi :_1 \bar{Q}/H_2 \to_2 \bar{Q}/H_2$$

which is constructed in terms of explicit coordinates on $Q/\Sigma(3)$, gets a more transparent interpretation later in §15.23.

In §12, we summarize the known facts about discrete Γ_μ.

In §13, we explain by an explicit example how Kummer-type identities [Kummer, 1836, Abschnitt III p. 77, Abschnitt IV p. 128] can be obtained for hypergeometric functions of n-variables by using Theorem 7.1, the generalization of Riemann's theorem to n-variables.

In §14 we introduce the notion of *orbifold*, a normal analytic space with distinguished *ramification divisors* to which are attached *weights* which are reciprocal integers. Our principal example of an orbifold is Q_μ^{st}/Σ in case μ is an S-uple satisfying condition ΣINT. We have occasion to consider structures which are obtained from orbifolds by formally dropping the condition that the weights be reciprocal integers (cf. 16.3, 17.13).

In §15, we extend the results in DM §13 for elliptic and euclidean μ replacing the INT condition assumed there by the weaker ΣINT hypothesis. The resulting larger class of orbifolds yield a number of the ball quotients constructed by Barthels-Hirzebruch-Hofer as ramified covers of blowups of the Hesse and extended Hesse line arrangement in $\mathbf{P}^2$ (cf. 15.6), and of line arrangements on abelian surfaces (cf. 15.19, 15.20, 15.22).

In §16 we show that the type III construction of R. Livne (On certain covers of the universal elliptic curve, Ph.D. Thesis, Harvard (1981)) is of the type Q_μ^{st} for suitable μ.

Finally in §17, we present definitions of χ_{orb}: orbifold Euler Poincaré invariants, and K_{orb}: the orbifold canonical class. This permits one to adapt the computations in Barthels-Hirzebruch-Hofer for their surface invariant $Prop(S) := 3\chi(S) - K^2$ to the case of orbifold, and to shed light on the construction of orbifolds with vanishing $Prop_{orb}$. Thereafter, we pose some questions about the complex differential geometry of families of formal orbifolds with varying weights, which if answerable positively, will permit the determination of the arithmeticity of some and hopefully all the remaining ball quotients constructed in Barthels-Hirzebruch-Hofer.

The authors have attempted to make this monograph intelligible to readers of diverse backgrounds.

1.N. Notations

We will systematically use the following notations:

S: a set with $N \geq 3$ elements; $n = N - 3$.

$\mathbf{P}^1$: the standard projective line $\mathbf{C} \cup \{\infty\}$. It is also $(\mathbf{C}^2 - \{0\})/\mathbf{C}^*$, with (z_1, z_2) mapping to z_1/z_2.

M: $\mathbf{P}^{1^S}$ minus the diagonals $x_s = x_t$. The group $PGL(2)$ acts on $\mathbf{P}^1$, hence on M. On M, the action is free.

Q: $M/PGL(2)$. The space Q is of complex dimension n.

For $N \geq 4$, we also consider

M^+: the space of S-uples x with at most one coincidence $x_s = x_t$. The group $PGL(2)$ acts freely on M^+.

Q^+: $M^+/PGL(2)$.

$D_{s,t}$: image in Q^+ of the divisor $x_s = x_t$ of M^+. It is a smooth divisor in Q^+, with inverse image in M^+ the smooth divisor $x_s = x_t$.

At times, we will fix a, b, c in S and put: $S_1 = S - \{a, b, c\}$,

M_1: the subset of M with $x_a = 0, x_b = 1, x_c = \infty$. One has $PGL(2) \times M_1 \xrightarrow{\sim} M$ and the quotient map induces an isomorphism $M_1 \xrightarrow{\sim} Q$. The projection to $\mathbf{P}^{1^{S_1}}$ identifies M_1 with $\mathbf{P}^{1^{S_1}}$ minus the divisors $x_s = 0, 1, \infty (s \in S_1)$ and $x_s = x_t$ $(s, t$ distinct in $S_1)$.

We write $\binom{S}{2}$ for the set of 2-elements subsets of S. We will have to consider families μ of complex numbers indexed by S, and families α (or β) of complex numbers indexed by $\binom{S}{2}$. We write $\alpha_{s,t}$ for $\alpha_{\{s,t\}}$: $\alpha_{s,t} = \alpha_{t,s}$ and $\alpha_{s,s}$ is undefined.

When, in an argument, S is to vary, we will write $M(S)$, $Q(S), \ldots$ instead of $M, Q, \ldots$.

§2. PICARD GROUP AND COHOMOLOGY

2.1. In this section, we will deal with a complex algebraic variety V, a scheme of finite type over $\mathbf{C}$; V or equivalently the corresponding analytic variety V^{an} is supposed to be smooth. We write $\mathcal{O}$ (resp. $\mathcal{O}^{an}$) for the structural sheaf of V (resp. V^{an}); the topology is the Zariski and the classical topology respectively. In particular, $\mathcal{O}(V)$ is the ring of regular functions on V, $\mathcal{O}^{an}(V)$ is the ring of holomorphic functions on V, and $\mathcal{O}^*(V)$ is the multiplicative group of invertible regular functions on V. Assume that V is separated and that $\bar{V}$ is a normal compactification of V. The ring $\mathcal{O}(V)$ can then be described analytically as the ring of holomorphic functions on V^{an} which are meromorphic at infinity (i.e., along $\bar{V} - V$).

2.2. Assume that V is connected and separated (V^{an} Hausdorff). Let $\bar{V}$ be a normal compactification of V and let $D_i(i \in I)$ be the irreducible codimension one subvarieties of $\bar{V}$ in $\bar{V} - V$. At a general point of D_i, $\bar{V}$ is non singular and D_i is defined by an equation $z_i = 0$. The *valuation* $v_{D_i}(f)$ of $f \in \mathcal{O}^*(V)$ along D_i is the integer n such that $z_i^{-n} f$ extends across D_i as an holomorphic invertible function, at a general point of D_i. A function $f \in \mathcal{O}^*(V)$ all of whose valuations are zero is holomorphic on $\bar{V}$ outside of a subvariety of codimension 2; hence it is a constant. The group $\mathcal{O}^*(V)/\mathbf{C}^*$ hence embeds into $\mathbf{Z}^I$; it is finitely generated.

PROPOSITION 2.3. *The following conditions are equivalent:*

(i) $Pic(V)$ is finitely generated.

(ii) There exists a non-empty Zariski-open subset W of V such that $Pic(W) = 0$.

PROOF. Let W be a non-empty Zariski-open subset of V and let $D_j(j \in J)$ be the irreducible codimension one subvarieties in $V - W$. The interpretation of Pic as the quotient of the group of divisors by the subgroup of principal divisors gives an exact sequence

$$(2.3.1) \quad 0 \to \mathcal{O}^*(V)/\mathbf{C}^* \to \mathcal{O}^*(W)/\mathbf{C}^* \xrightarrow{v} \mathbf{Z}^J \xrightarrow{D} Pic(V) \to Pic(W) \to 0$$

with $v : f \mapsto (v_{D_j}(f))_{j \in J}$ and $D : (x_j) \mapsto \Sigma x_j D_j$.

(i) $\Rightarrow$ (ii): if the effective divisors D_i span $Pic(V)$, one may take for W the complement of the D_i.

(ii) $\Rightarrow$ (i): If $Pic(W) = 0$, then $\mathbf{Z}^J$ maps onto $Pic(V)$.

The following is a particular case of (2.3.1).

COROLLARY 2.4. *If $H^0(V, \mathcal{O}^*)$ is reduced to the constants and $PicW = 0$, one has*

$$0 \to \mathcal{O}^*(W)/\mathbf{C}^* \xrightarrow{v} \mathbf{Z}^J \to Pic(V) \to 0$$

is exact.

2.5. The exponential sequence of sheaves on V^{an}:

$$(2.5.1) \qquad\qquad 0 \to 2\pi i \mathbf{Z} \to \mathcal{O}^{an} \to \mathcal{O}^{an^*} \to 0$$

defines coboundary maps which, composed with the pullback by $V^{an} \to V$, yield maps

$$\mathcal{O}^*(V) \to \mathcal{O}^*(V^{an}) \xrightarrow{\partial} H^1(V, \mathbf{Z})$$
$$(2.5.2) \qquad\qquad Pic(V) \to H^1(V^{an}, \mathcal{O}^*) \xrightarrow{\partial} H^2(V, \mathbf{Z}),$$

where $H^i(V, \mathbf{Z})$ is the transcendental cohomology of V^{an}.

PROPOSITION 2.6. *If $Pic(V)$ is finitely generated (cf. 2.3), then $\mathcal{O}^*(V)/\mathbf{C}^*$ maps isomorphically to $H^1(V, \mathbf{Z})$ and $Pic(V)$ injects into $H^2(V, \mathbf{Z})$, with a torsion free cokernel.*

PROOF. Let μ_n be the group of nth roots of unity. We will use that $Pic(V)$ can be interpreted as $H^1(V, \mathcal{O}^*)$, for the etale homology of V (SGA4 IX 3.3). This results from its interpretation as the set of isomorphism classes of line bundles on V, and etale descent. We will also use that the etale cohomology groups $H^i(V, \mu_n)$ coincide with the classical $H^i(V^{an}, \mu_n)$ (SGA 4 XI).

The Kummer exact sequence on V for the etale topology

$$0 \to \mu_n \to \mathcal{O}^* \xrightarrow[f \mapsto f^n]{} \mathcal{O}^* \to 0$$

yields a long exact sequence of etale cohomology groups
(2.6.1)
$$0 \to \mu_n \to \mathcal{O}^*(V) \to \mathcal{O}^*(V) \to H^1(V, \mu_n) \to Pic(V) \to Pic(V) \to H^2(V, \mu_n)$$

having as quotient the long exact sequence

$$0 \to \mathcal{O}^*(V)/\mathbf{C}^* \to \mathcal{O}^*(V)/\mathbf{C}^* \to H^1(V, \mu_n) \to Pic(V) \to Pic(V) \to H^2(V, \mu_n).$$

This sequence can be compared with the long exact sequence deduced from the short exact sequence on V^{an}:

$$0 \to \mathbf{Z} \xrightarrow{n} \mathbf{Z} \to \mathbf{Z}/(n) \to 0,$$

where $\mathbf{Z}/(n)$ is to be identified with μ_n by $a \mapsto exp(2\pi i a/n)$: one has a morphism of long exact sequences
(2.6.2)
$$
\begin{array}{ccccccccccc}
0 \to & \mathcal{O}^*(V)/\mathbf{C}^* & \to & \mathcal{O}^*(V)/\mathbf{C}^* & \to & H^1(V,\mu_n) & \to & Pic(V) & \to & Pic(V) & \to & H^2(V,\mu_n) \\
 & \downarrow & & \downarrow & & \| & & \downarrow & & \downarrow & & \| \\
0 \to & H^1(V,\mathbf{Z}) & \to & H^1(V,\mathbf{Z}) & \to & H^1(V,\mathbf{Z}/(n)) & \to & H^2(V,\mathbf{Z}) & \to & H^2(V,\mathbf{Z}) & \to & H^2(V,\mathbf{Z}/(n))
\end{array}
$$

The proof is given below. It results from (2.6.1) and (2.6.2) that an element of $\mathcal{O}^*(V)/\mathbf{C}^*$ or $Pic(V)$, with trivial image in $H^1(V, \mathbf{Z})$ or $H^2(V, \mathbf{Z})$, is divisible by n. This holds for all n, hence such an element is zero, and the vertical maps (2.6.2) are injective.

As $Pic(V) \to H^2(V, \mathbf{Z})$ is injective, an element of $H^1(V, \mathbf{Z})$ has an image in $H^1(V, \mathbf{Z}/(n)) = H^1(V, \mu_n)$ coming from $\mathcal{O}^*(V)/\mathbf{C}^*$; i.e. $\mathcal{O}^*(V)/\mathbf{C}^*$ maps onto $H^1(V, \mathbf{Z})/nH^1(V, \mathbf{Z})$. Since this holds for all n,

$$\mathcal{O}^*(V)/\mathbf{C}^* \to H^1(V, \mathbf{Z}) \text{ is surjective.}$$

If an element of $H^2(V, \mathbf{Z})$ is of n-torsion, it is in the image of $H^1(V, \mathbf{Z}/(n))$, hence in that of $Pic(V)$. The cokernel of $Pic(V) \to H^2(V, \mathbf{Z})$ is hence torsion free.

The compatibility (2.6.2) can be proved as follows.

The long exact sequence deduced from the Kummer short exact sequence of sheaves maps to the similarly defined one on V^{an}. The arguments now take place on V^{an}. The cohomology of μ_n is the same as the hypercohomology of the complex in degree 0 and 1 $[\mathcal{O}^* \xrightarrow{\nu} \mathcal{O}^*]_{0,1}$ where ν denotes $f \to f^n$, and the long exact sequence is deduced from maps of complexes (a distinguished triangle, cf. J.L. Verdier, Catégories dérivées, SGA $4\frac{1}{2}$)

(2.6.3)

$$[\mathcal{O}^* \xrightarrow{\nu} \mathcal{O}^*]_{0,1} \to [\mathcal{O}^*]_0 \xrightarrow{\nu} [\mathcal{O}^*]_0$$

$$\to [\mathcal{O}^* \xrightarrow{\nu} \mathcal{O}^*]_{-1,0} \to [\mathcal{O}^*]_{-1} \xrightarrow{\nu} [\mathcal{O}^*]_{-1} \to [\mathcal{O}^* \xrightarrow{\nu} \mathcal{O}^*]_{-2,-1}.$$

The complex $[\mathcal{O}^*]_0$ reduced to $\mathcal{O}^*$ in degree zero maps to $[\mathcal{O} \xrightarrow{exp} \mathcal{O}^*]_{-1,0}$, quasi-isomorphic to $\mathbf{Z}$ in degree -1. This allows us to go from (2.6.3) to

$$\begin{bmatrix} \mathcal{O}^* \to & \mathcal{O}^* \\ \uparrow & \uparrow \\ \mathcal{O} \to & \mathcal{O} \end{bmatrix} \to \begin{bmatrix} \mathcal{O}^* \\ \uparrow \\ \mathcal{O} \end{bmatrix} \to \begin{bmatrix} \mathcal{O}^* \\ \uparrow \\ \mathcal{O} \end{bmatrix} \to \begin{bmatrix} \mathcal{O}^* \to & \mathcal{O}^* \\ \uparrow & \uparrow \\ \mathcal{O} \to & \mathcal{O} \end{bmatrix} \to \begin{bmatrix} \mathcal{O}^* \\ \uparrow \\ \mathcal{O} \end{bmatrix} \to \cdots$$

with the same hypercohomology as

$$[\mathbf{Z} \to \mathbf{Z}]_{-1,0} \to [\mathbf{Z}]_{-1} \to [\mathbf{Z}]_{-1} \to [\mathbf{Z} \to \mathbf{Z}]_{-2,-1} \to [\mathbf{Z}]_{-2} \to \cdots$$

which gives the second line of (2.6.2). We have been careless with sign questions, but this is irrelevant for the use we made of (2.6.2).

2.7. The exact sequence (2.3.1) is compatible with the Gysin sequence

$$0 \to H^1(V, \mathbf{Z}) \to H^1(W, \mathbf{Z}) \to \mathbf{Z}^J \to H^2(V, \mathbf{Z}) \to H^2(W, \mathbf{Z})$$

and the monomorphisms of Proposition 2.6.

2.8. Fix $W \subset V$ as in 2.4. Let us consider the multiplicative group T of multivalued functions f on W of the form $f = \Pi f_k^{\alpha_k}$ with $f_k \in \mathcal{O}^*(W)$ and $\alpha_k \in \mathbf{C}$. We view them as functions on a fixed universal covering of W. At a general point of D_i, if $z = 0$ is a local equation for D_i, f can be written near D_i as $f = z^v \cdot h$ with h holomorphic invertible. We call v the valuation $v_i(f)$ of f along D_i. One has

$$v_i(f) = \Sigma \alpha_k v_i(f_k).$$

The diagram deduced from 2.4

$$(\mathcal{O}^*(W)/\mathbf{C}^*) \otimes \mathbf{C} \xrightarrow{v \otimes \mathbf{C}} \mathbf{C}^I$$

$$\searrow \qquad \qquad \nearrow v$$

$$T/\mathbf{C}^*$$

shows that

$$(\mathcal{O}^*(W)/\mathbf{C}^*) \otimes \mathbf{C} \xrightarrow{\sim} T/\mathbf{C}^*.$$

We hence have an exact sequence

$$0 \to T/\mathbf{C}^* \to \mathbf{C}^J \to Pic(V) \otimes \mathbf{C} \to 0.$$

Remark. The isomorphism $\mathcal{O}^*(W)/\mathbf{C}^* \to H^1(W, \mathbf{Z})$ induces a map

$$(\mathcal{O}^*(W)/\mathbf{C}^*) \otimes \mathbf{C} \xrightarrow{\sim} H^1(W, \mathbf{C}) \xrightarrow[\exp(2\pi i z)]{} H^1(W, \mathbf{C}^*) = Hom(\pi_1(W), \mathbf{C}^*).$$

To $\Sigma f_k \otimes \alpha_k$ it attaches the monodromy of $\Pi f_k^{\alpha_k}$. As $Pic(W) = 0$, (cf. the remark following 2.6.2) $H^2(W, \mathbf{Z})$ is torsion free and $H^1(W, \mathbf{C}^*) = H^1(W, \mathbf{Z}) \otimes \mathbf{C}^*$. The above map hence induces an isomorphism

$$T/\mathcal{O}^*(W) \xrightarrow{\sim} Hom(\pi_1(W), \mathbf{C}^*).$$

2.9. Let V be a non-singular algebraic variety, W be a Zariski open dense subset of V, $\mathcal{L}$ be a line bundle on V and $\nabla : \mathcal{L} \to \Omega^1 \otimes \mathcal{L}$ an integrable connection on the restriction of $\mathcal{L}$ to W. Let D_i $(i \in I)$ be the irreducible components of $V - W$ which are of codimension one in V.

Let g be a local trivialization of $\mathcal{L}$ at a general point of D_i. We assume that the 1-form $\frac{\nabla g}{g}$ has a simple pole along each D_i; by the hypothesis on ∇, $\frac{\nabla g}{g}$ is holomorphic on W. It is a closed 1-form and its residue is a constant on D_i. The residue is independent of the choice of g and we call it the residue of ∇ along D_i. When $\mathcal{L}$ is replaced by $\mathcal{L}(nD_i)$, the residue is changed by the addition of $-n$.

PROPOSITION 2.10. *The first Chern class of $\mathcal{L}$ in*

$$H^2(V, \mathbf{C}) = H^2(V, \mathbf{Z}) \otimes \mathbf{C}$$

is given by

(2.10.1) $$c_1(\mathcal{L}) = -\Sigma \; Res_{D_i}(\nabla) \; c\ell(D_i)$$

A related statement is given in [Esnault and Viehweg, 1986] App. B, where vector bundles with logarithmic connections and Chern classes c_p are considered. The proof there can be modified to yield 2.10 but we do not simply quote that paper, because Chern classes there are taken in $H^p(V, \Omega^p)$, related to $H^2(V, \mathbf{C})$ by Hodge theory only on a compact Kaehler variety.

In 2.10, the assumption that V (and $\mathcal{L}$) is algebraic is not necessary, but permits the following proof.

PROOF. We may remove from V subspaces of codimension 2: this does not change $H^2(V, \mathbf{Z})$. We may shrink W: this does not change the second member of (2.10.1) since, by hypothesis, $\frac{\nabla g}{g}$ is a holomorphic 1-form if g is a trivialization of $\mathcal{L}$ around a point of W. We hence may and shall assume that $\mathcal{L}$ is trivial on W and that the D_i are disjoint smooth divisors with union $V - W$.

A trivialization of $\mathcal{L}$ on W extends to an isomorphism $\mathcal{L} = \mathcal{O}(\Sigma\, n_i D_i)$. Replacing $\mathcal{L}$ by $\mathcal{L}(-\Sigma n_i D_i)$ adds to both sides of (2.10.1) the sum $-\Sigma n_i c\ell(D_i)$: we may and shall assume that $\mathcal{L} = \mathcal{O}$. The connection ∇ is then given by a closed 1-form ν with logarithmic poles along the D_i where:

$$(2.10.2) \qquad\qquad \nabla = d - \nu$$

Thus $Res_{D_i}(\nabla) = -Res_{D_i}(\nu)$. Formula (2.10.1) reduces to the lemma:

LEMMA 2.11. *If ν is a closed 1-form with logarithmic poles along smooth divisors D_i on a complex manifold V, one has in $H^2(V, \mathbf{C})$:*

$$\Sigma\, Res_{D_i}(\nu)c\ell(D_i) = 0.$$

PROOF. A function with a simple pole along a smooth divisor is locally L^1, hence is a generalized function (= distribution). Similarly, a 1-form with simple pole ν defines a 1-form with generalized functions as coefficients, denoted $v.p.\nu$. If ν is closed, $v.p.\nu$ need not be closed: its exterior derivative, a 2-form with generalized functions as coefficients, is given by

$$(2.11.1) \quad \frac{1}{2\pi i}d\, v.p.\nu = \Sigma\, Res(\nu)[D_i], \quad (\text{cf. [Schwartz, 1957] II.3.28, p. 49})$$

where $[D_i]$ is the integration current over D_i. At least if V is Hausdorff, the complex cohomology can be computed using the de Rham complex with generalized functions coefficients, and the class of D_i is represented by the integration current over D_i. Formula (2.11.1) then proves 2.11. Even if V is not Hausdorff, the cohomology of the de Rham complex maps to ordinary cohomology and this suffices for the argument.

2.12. We will use 2.10 in conjunction with 2.6, interpreting an equality in $H^2(V, \mathbf{Z}) \otimes \mathbf{C}$ as an equality in $Pic(V) \otimes \mathbf{C}$.

§3. COMPUTATIONS FOR Q AND Q^+

3.1. Fix a set S with $N \geq 4$ elements. Define M, Q, M^+, Q^+ and $D_{s,t}$ as in §1, N. We will compute the functors $\mathcal{O}^*$ and Pic on Q and Q^+.

For s and t distinct in S, the valuation along the smooth divisor $D_{s,t} \subset Q^+$ is a map

$$(3.1.1) \qquad\qquad v_{s,t} : \mathcal{O}^*(Q) \to \mathbf{Z}.$$

If $\binom{S}{2}$ is the set of 2-elements subsets of S, the maps (3.1.1) define

$$(3.1.2) \qquad\qquad v : \mathcal{O}^*(Q) \to \mathbf{Z}^{\binom{S}{2}}.$$

One defines a map

$$(3.1.3) \qquad\qquad \mathbf{Z}^{\binom{S}{2}} \to Pic(Q^+)$$

by sending each basis element to the class of corresponding divisor $D_{s,t}$.

Fix $s \in S$ and fix distinct elements $x_t^0 \in \mathbf{P}^1$ ($t \in S - \{s\}$). The $x \in M^+$ with $x_t = x_t^0$ for $t \neq s$ form a projective line. For any line bundle $\mathcal{L}$ on Q^+, let d_s be the degree of the pull back of $\mathcal{L}$ on the projective line; it defines

$$(3.1.4) \qquad\qquad d_s : Pic(Q^+) \to \mathbf{Z}$$
$$(3.1.5) \qquad\qquad d : Pic(Q^+) \to \mathbf{Z}^S.$$

One has $d_s(D_{i,j}) = 1$ for $s = i$ or j, 0 otherwise. The composite of (3.1.5) and (3.1.3) is

$$(3.1.6) \qquad T : \mathbf{Z}^{\binom{S}{2}} \to \mathbf{Z}^S : \mathbf{n} \mapsto \left(\sum_{t \neq s} n_{s,t} \right)_{s \in S}.$$

PROPOSITION 3.2.

(i) $\mathcal{O}^*(Q^+) = \mathbf{C}^*$.

(ii) *(3.1.5) induces an isomorphism* $Pic(Q^+) \xrightarrow{\sim} \ker(\Sigma : \mathbf{Z}^S \to \mathbf{Z}/(2))$.

(iii) *The sequence*

$$0 \to \frac{\mathcal{O}^*(Q)}{\mathbf{C}^*} \xrightarrow{v} \mathbf{Z}^{\binom{S}{2}} \xrightarrow{T} \mathbf{Z}^S \to \mathbf{Z}/(2) \to 0$$

is exact.

(iv) $Pic(Q) = 0$.

Proof of (i) Since Q^+ is a quotient of M^+, it suffices to see that $\mathcal{O}(M^+) = \mathbf{C}$. The open subset $M^+ \subset \mathbf{P}^{1^S}$ is the complement of a subset of codimension ≥ 2 of $\mathbf{P}^{1^S}$. Therefore any holomorphic function on M^+ extends to $\mathbf{P}^{1^S}$ and is constant.

Proof of (iv) Fix $a, b, c \in S$ and define S_1, M_1 as in §1N. The subspace M_1 of M maps isomorphically to Q. It can be identified with the complement in the affine space A^{S_1} of the hyperplanes $x_i = 0, x_i = 1$ ($i \in S_1$) and $x_i = x_j$. The closure in A^{S_1} of any divisor D of M_1 is a divisor of A^{S_1}, automatically principal; hence $Pic(M_1) = Pic(Q) = 0$.

Proof of (ii)(iii) By (2.4), one has

(3.2.1) $0 \to \mathcal{O}^*(Q)/\mathbf{C}^* \xrightarrow{v} \mathbf{Z}^{\binom{S}{2}} \to Pic(Q^+) \to 0$

is exact.

For i, j, k, ℓ distinct in S, the cross ratio $(x_i, x_j, x_k, x_\ell) : M^+ \to \mathbf{P}^1$ descends to a function from Q^+ to $\mathbf{P}^1$. The inverse image of 0 (resp. ∞) is $D_{ik} + D_{j\ell}$ (resp. $D_{jk} + D_{i\ell}$). The cross ratio is an invertible function on Q whose image by v is $e_{ik} + e_{j\ell} - e_{jk} - e_{i\ell}$. Let R be the subgroup of $\mathbf{Z}^{\binom{S}{2}}$ generated by these images. Diagram chasing in

$$\begin{array}{ccccccc}
0 \to & \mathcal{O}^*(Q)/\mathbf{C}^* & \to & \mathbf{Z}^{\binom{S}{2}} & \to & PicQ^+ & \to 0 \\
& \cup & & & & d \downarrow & \\
& R & & \xrightarrow{T} & \mathbf{Z}^S & \xrightarrow{\Sigma} \mathbf{Z}/(2), &
\end{array}$$

we see that to prove (ii), (iii) and that constants and cross ratios generate $\mathcal{O}^*(Q)$, it is enough to check:

LEMMA 3.3. *For $|S| \geq 3$, the sequence*

$$R \to \mathbf{Z}^{\binom{S}{2}} \xrightarrow{T} \mathbf{Z}^S \to \mathbf{Z}/(2) \to 0$$

is exact.

PROOF. The cokernel of T is generated by elements e_s ($s \in S$), with relations $e_s = -e_t$ for $s \neq t$. For s, t, u distinct, this implies $e_s = -e_t = e_u = -e_s$. One infers $2e_s = 0$ and $e_s = e_t$: the cokernel is $\mathbf{Z}/(2)$.

We prove exactness at $\mathbf{Z}^{\binom{S}{2}}$ by induction on $|S| \geq 3$. If $|S| = 3$, $R = 0$ and T is injective. If $|S| \geq 4$, fix $s_0 \in S$. If $\mathbf{a}$ lies in the kernel of T, one has $\sum_t a_{s_0, t} = 0$. One can add to $\mathbf{a}$ an integral linear combinations of the generators of R and replace $\mathbf{a}$ by an element in the kernel with $a_{s_0, t} = 0$ for all $t \neq s_0$ in S. This reduces the problem to the case of $S - \{s_0\}$ and proves 3.3 by induction.

For later use, we mention the corollary:

COROLLARY 3.4. *Fix $s \in S$. The Picard group of $Q^+ - \bigcup_t D_{s,t}$ is $\mathbf{Z}$. All divisors $D_{k,\ell}$ ($k, \ell \neq s$) are linearly equivalent and their class is a generator of Pic.*

PROOF. The Picard group is the quotient of $Pic(Q^+) = \ker(\mathbf{Z}^S \to \mathbf{Z}/(2))$ by the subgroup generated by the $e_s + e_t$ with s fixed, t varying. The map $\mathbf{Z}^S \to \mathbf{Z} : \mathbf{a} \mapsto (\sum_{t \neq s} a_t) - a_s$ identifies the quotient with $2\mathbf{Z}$, and 3.4 follows.

Applying 2.8, we get

COROLLARY 3.5. *Multivalued functions on Q of the form $f = \prod f_k^{\alpha_k}$ with $f_k \in \mathcal{O}^*(Q)$ and $\alpha_i \in \mathbf{C}$ are uniquely determined, up to a multiplicative constant, by their valuations $v_{s,t}(f) = \Sigma \alpha_i v_{s,t}(f_i)$. The valuations can be any system of complex numbers $\mathbf{v} \in \mathbf{C}^{\binom{S}{2}}$ for which $\sum_{t:(t \neq s)} v_{s,t} = 0$ for each $s \in S$.*

Remarks 3.6 (i) The fact that the constants and the cross-ratios generate $\mathcal{O}^*(Q)$ is also easily seen on the model $Q \simeq M_1 \subset A^{S_1}$: the functions $x_i, 1 - x_i$ and $(x_i - x_j)/x_i$ are among the cross-ratios.

(ii) The same model shows that $f \in \mathcal{O}^*(Q)$ is determined up to constants by $v_{a,i}(f), v_{b,i}(f)$ and $v_{i,j}(f)$ $(i, j \in S_1)$, and that these valuations can be freely chosen.

(iii) A more geometric proof of $Pic(Q^+) = \{\mathbf{a} \in \mathbf{Z}^S \mid \Sigma a_s \in 2\mathbf{Z}\}$ will be given in (3.14).

3.7. The case $N = 4$ For $N = 4$, the space Q^+ is not separated: the cross ratio maps it to $\mathbf{P}^1$, but the points $0, 1, \infty$ have to be doubled. Each corresponds to a partition $\{\{s, t\}\{u, v\}\}$ of S, and the two points above are given respectively by $x_s = x_t$ and $x_u = x_v$. The subspace in 3.4 is just a projective line: one point above each of $0, 1, \infty$ is deleted.

PROPOSITION 3.8. *The class in $Pic(Q^+) \subset \mathbf{Z}^S$ of the maximal exterior power of $\Omega^1_{Q^+}$ is $(-2, \ldots, -2)$.*

PROOF. One has to show that for each $s \in S$, one has $d_s(\Omega^{\max}_{Q^+}) = -2$. Fix $a, b, c \neq s$ in S and define $S_1 = S - \{a, b, c\}$. Fix distinct $x_t^0 \in \mathbf{P}^1$ for $t \neq s$, with $x_a^0 = 0, x_b^0 = 1, x_c^0 = \infty$. The d_s to be evaluated is a degree on the projective line $x_t = x_t^0$ for $t \neq s$. This line embeds in $Q^+ - D_{ab} - D_{bc} - D_{ca}$, which can be identified with the open subset of $\mathbf{P}^{S_1}$ obtained by removing the intersections, two by two, of the divisors $x_{t_1} = 0, x_{t_2} = 1, x_{t_3} = \infty$ and $x_s = x_t$. The embedded line is obtained by fixing all coordinates but one in $\mathbf{P}^{S_1}$, and since the restriction of the canonical bundle of $\mathbf{P}^{S_1}$ to this line is its Ω^1,

$$d_s\left(\Omega^{\max}_{Q^+}\right) = \deg(\mathbf{P}^1, \Omega^1) = -2.$$

COROLLARY 3.9. *The class in $Pic(Q^+)$ of the maximal exterior power of $\Omega^1_{Q^+}(\log \Sigma D_{s,t})$ is $(N - 3, \ldots, N - 3)$.*

PROOF. The maximal exterior power is

$$\Omega^{\max}_{Q^+}(\Sigma D_{s,t}) \qquad \text{(sum over } \binom{S}{2}\text{)}.$$

Each degree d_s is $d_s\left(\Omega^{\max}_{Q^+}\right) + \sum_{t \neq s} 1 = -2 + (N - 1) = N - 3$.

3.10 Let S_+ be deduced from S by the adjunction of an element w and let

(3.10.1) $$\pi : Q^+(S_+) \to Q^+(S)$$

be the "forgetting x_w"-map. A point p in $Q^+(S)$ is an isomorphism class of systems $(P, (x_s)_{s \in S})$: P a projective line, (x_s) a system of points with at most one confluence $x_s = x_t$. The map (3.10.1) is smooth, with fiber at p the projective line P if the x_s are distinct, and P minus the x_s if there is a confluence $x_s = x_t$. For $\mathcal{L}$ a line bundle on $Q^+(S_+)$, the degree d_w is the degree of the restriction of $\mathcal{L}$ to a general fiber of (3.10.1). For $s \neq w$, the degree d_s on $Q^+(S)$ and $Q^+(S_+)$ is the degree on suitable projective lines. The one for $Q^+(S)$ is the image of the one for $Q^+(S_+)$. The natural inclusion of $\mathbf{Z}^S$ into $\mathbf{Z}^{S_+}$ hence gives rise to a commutative diagram

(3.10.2)
$$
\begin{array}{ccc}
Pic(Q^+(S)) & \longrightarrow & Pic(Q^+(S_+)) \\
\| & & \| \\
\ker(\mathbf{Z}^S \to \mathbf{Z}/(2)) & \longrightarrow & \ker(\mathbf{Z}^S \to \mathbf{Z}/(2)).
\end{array}
$$

COROLLARY 3.11. *The class in $Pic(Q^+(S_+))$ of $\Omega^1_{Q^+(S_+)/Q^+(S)}$ is given by $d_w = -2$ and $d_s = 0$ for $s \neq w$.*

PROOF. The line bundle in question is

$$\Omega^{\max}(Q^+(S_+)) \otimes \pi^* \Omega^{\max}(Q^+(S))^{-1}$$

and one applies 3.9, 3.10.2.

3.12 In the remainder of this paragraph, we give a second proof, inspired by Gelfand et al. [cf. collected works], of the isomorphism

$$Pic(Q^+) = \ker(\mathbf{Z}^S \to \mathbf{Z}/(2)).$$

The map $H^+ \to Q^+$ at the core of the argument will reappear in §5.

We define

$$\bar{H} := \hom(\mathbf{C}^2, \mathbf{C}^S);$$

if we take $S = \{1, \ldots, N\}$, it is the vector space of $N \times 2$ matrices. The space $\bar{H}$ has coordinates $h_{s,i}$ with $i = 1, 2$ and $s \in S$. For $h \in H$, we denote by h_s the linear form $pr_s \circ h$ on $\mathbf{C}^2$; its coordinates are $(h_{s,1}, h_{s,2})$. If $h_s \neq 0$, we let $h_s^{\perp}$ be the point of $\mathbf{P}^1$ defined by the line $h_s = 0$; its projective coordinates are $(h_{s,2}, -h_{s,1})$.

For a, b distinct in S, we define

$$F_{a,b}(h) := \det \begin{pmatrix} h_{a,1} & h_{a,2} \\ h_{b,1} & h_{b,1} \end{pmatrix}.$$

The non-vanishing $F_{a,b}(h) \neq 0$ means that h_a and h_b are linearly independent; this holds if and only if $h_a \neq 0, h_b \neq 0$ and $h_a^{\perp} \neq h_b^{\perp}$.

Let H be the complement in $\bar{H}$ of the divisors $F_{a,b} = 0$.

We have a map

(3.12.1) $$H \to M \subset \mathbf{P}^{1^S} : h \mapsto (h_s^{\perp}).$$

The group $\mathbf{C}^{*S}$ acts on $\bar{H}$ by left composition with a diagonal matrix:

$$(\lambda h)_s = \lambda_s h_s \qquad (s \in S)$$

and H is a $\mathbf{C}^{*S}$-torsor over M, i.e. H is a principal $\mathbf{C}^{*S}$ bundle over M.

The group $SL(2)$ acts on $\bar{H}$ by $h \mapsto h \circ g^{-1}$. The action of $-1 \in SL(2)$ coincides with the action of $-1 := (-1, \ldots, -1) \in \mathbf{C}^{*S}$. On H, we get a free action of

$$(\mathbf{C}^{*S} \times SL(2))/\{(1,1),(-1,-1)\}.$$

Dividing by $SL(2)$, we get that

(3.12.2) $$H/SL(2) \to Q = M/PGL(2)$$

is a $\mathbf{C}^{*S}/\{\pm 1\}$-torsor.

Let $H^+ \subset \bar{H}$ be the open subset of $\bar{H}$ consisting of the h with $h_s \neq 0$ ($s \in S$) and at most one of the $F_{a,b}$ vanishing. The complement of H^+ in $\bar{H}$ is of codimension 2. For the scheme H^+, one has therefore

$$\mathcal{O}^*(H^+) = \mathcal{O}^*(\bar{H}) = \mathbf{C}^*$$
(3.12.3) $$Pic(H^+) = Pic(\bar{H}) = 0.$$

We recall that $M^+ \subset \mathbf{P}^S$ is the space of x with at most one coincidence $x_a = x_b$, that Q^+ is $M^+/PGL(2)$, and that $D_{a,b} \subset Q^+$ is the image of the

divisor $x_a = x_b$ of M^+. On H^+, the action $\mathbf{C}^{*S} \times SL(2)/\{(1,1)(-1,-1)\}$
is free, H^+ is a $\mathbf{C}^{*S}$-torsor over M^+ and, dividing by $SL(2)$, one gets that

(3.12.4)
$$H^+/SL(2) \to Q^+$$

is a $\mathbf{C}^{*S}/\{\pm1\}$-torsor. The functions $F_{a,b}$ on $\bar{H}$ are $SL(2)$-invariant, hence
descend as functions on $H^+/SL(2)$; their divisors of zeros are the inverse
images of the divisors $D_{a,b}$ of Q^+.

We shall make use of the following well-known lemma.

LEMMA 3.13. *Let Y be a connected regular scheme and $f : X \to Y$ a
smooth morphism with (non-empty) connected fibers. Let η be the generic
point of Y: $k(\eta)$ is the field of rational functions on Y. Let X_η denote the
fiber of X at η. One has an exact sequence*

$$0 \to \mathcal{O}^*(Y) \to \mathcal{O}^*(X) \to \mathcal{O}^*(X_\eta)/k(\eta)^* \to Pic(Y) \to Pic(X) \to Pic(X_\eta) \to 0.$$

PROOF. If a rational function of X is regular invertible on X_η, it is
regular invertible on $f^{-1}(V)$ for a suitably small non-empty open subset V
of Y, and its divisor has support in $X - f^{-1}(V)$. Similarly, if a line bundle
on X is trivial on X_η, it is trivial on $f^{-1}(V)$ for a suitably small non-empty
open subset V of Y, and its class in Pic is the image of a divisor with
support in $X - f^{-1}(V)$. Such a divisor is the inverse image of a divisor in
Y with support in $Y - V$, and we get a morphism of long exact sequences

$$
\begin{array}{ccccccccccc}
0 & \to & \mathcal{O}^*(Y) & \to & k(\eta)^* & \to & \text{Div}(Y) & \to & Pic(Y) & \to & 0 \\
& & \downarrow & & \downarrow & & \| & & \downarrow & & \downarrow \\
0 & \to & \mathcal{O}^*(X) & \to & \mathcal{O}^*(X_\eta) & \to & \text{Div}(Y) & \to & Pic(X) & \to & Pic(X_\eta) & \to & 0
\end{array}
$$

from which 3.13 follows by diagram chasing.

Most of the chase is of a kind encountered in any abelian category: maps
$A \xrightarrow{f} B \xrightarrow{g} C$ give rise to a long exact sequence

$$0 \to \ker(f) \to \ker(gf) \to \ker(g) \to \text{coker}(f) \to \text{coker}(gf) \to \text{coker}(g) \to 0.$$

Applying this to the maps $k(\eta)^* \to \mathcal{O}^*(X_\eta) \to \text{Div } Y$, one obtains the
latter part of the exact sequence of the Lemma, but shortened to end as
$\ldots Pic(Y) \to \ker(Pic(X) \to Pic(X_\eta))$.

COROLLARY 3.13.1. *With the assumptions of 3.13, if for both the functors $T = \mathcal{O}^*(\ )$ and $T = Pic$ one has at the generic point η of Y $T(\eta) \xrightarrow{\sim} T(X_\eta)$, then $T(Y) \xrightarrow{\sim} T(X)$.*

Inasmuch as $Pic(\eta) = 0$, the hypotheses amount to:
$\mathcal{O}^*(X_\eta)/k(\eta)^* = 0$, $Pic(X_\eta) = 0$, and the conclusion is immediate from 3.13.

The corollary applies if X_η is an affine line. It also applies if X is a $SL(2)$)-torsor: one has to check that for $SL(2)$ over a field K, $T(K) \xrightarrow{\sim} T(SL(2))$ which can be seen as follows. Let H be the subgroup $\begin{pmatrix} 1 & * \\ 0 & 1 \end{pmatrix}$. The quotient $SL(2)/H$ is the affine plane minus the origin. Since the point 0 is of codimension 2, one has

$$T(A^2 - \{0\}) \xleftarrow{\sim} T(A^2) \xleftarrow{\sim} T(K).$$

In turn, $SL(2)$ is an affine-line-bundle over $SL(2)/H$, and $T(A^2 - \{0\}) \xrightarrow{\sim} T(SL(2))$. We have thus proved:

COROLLARY 3.13.2. *If X is a $SL(2)$-torsor on a connected regular scheme Y, one has $\mathcal{O}^*(Y) \xrightarrow{\sim} \mathcal{O}^*(X)$ and $Pic(Y) \xrightarrow{\sim} Pic(X)$.*

From 3.13 we can infer

COROLLARY 3.13.3. *If X is a G-torsor on a connected regular scheme Y, with G isomorphic to $\mathbf{G}_m^n$, with character group $X(G) = \mathbf{Z}^n$, one has a long exact sequence*

$$0 \to \mathcal{O}^*(Y) \to \mathcal{O}^*(X) \to X(G) \to Pic(Y) \to Pic(X) \to 0.$$

This follows from

$$(K[T_1, T_1^{-1}, \ldots, T_n, T_n^{-1}])^* \simeq K^* \times \mathbf{Z}^n \qquad \text{and}$$
$$Pic\, K[T_1, T_1^{-1}, \ldots, T_n, T_n^{-1}] = 0$$

In 3.13.3, the map $\mathcal{O}^*(X) \to X(G)$ is as follows: for any local trivialization of the G-torsor X, one attaches to $\varphi \in \mathcal{O}^*(X)$ the character χ such that $\varphi(y, g) = \varphi(y, e)\chi(g)$. The map $X(G) \to Pic(Y)$ attaches to χ the class in $Pic(Y)$ of the $\mathbf{G}_m$-torsor $\chi(X)$ deduced from X by $\chi : G \to \mathbf{G}_m$.

3.14 From (3.13.2) and (3.12.3), one gets

$$\mathcal{O}^*(H^+/SL(2)) = \mathbf{C}^*$$

(3.14.1) $$Pic(H^+/SL(2)) = 0.$$

Applying (3.12.4)(3.13.3) gives

$$\mathcal{O}^*(Q^+) = \mathbf{C}^*$$
$$Pic(Q^+) = X(\mathbf{G}_m^S/\{\pm 1\})$$

(3.14.2) $$= ker(\mathbf{Z}^S \xrightarrow{\Sigma} \mathbf{Z}/(2)).$$

For H, one has $Pic(H) = 0$ and $\mathcal{O}^*(H)$ is the product of $\mathbf{C}^*$ by the free abelian group with basis $\{F_{a,b}; a, b \in S \text{ and } a < b \text{ for any fixed ordering of } S\}$:

$$0 \to \mathbf{C}^* \to \mathcal{O}^*(H) \to \mathbf{Z}^{\binom{S}{2}} \to 0;$$

the map $\mathcal{O}^*(H) = \mathcal{O}^*(H/SL(2))$ to $X(\mathbf{G}_m^S/\{\pm 1\}) = ker(\mathbf{Z}^S \to \mathbf{Z}/(2))$ attaches to F the element $(d_s) \in \mathbf{Z}^S$, with d_s the degree of F in (x_s, y_s). We have $Pic(Q) = 0$. The sequence (3.13.3) hence gives

(3.14.3) $$0 \to \frac{\mathcal{O}^*(Q)}{\mathbf{C}^*} \to \mathbf{Z}^{\binom{S}{2}} \to ker(\mathbf{Z}^S \to \mathbf{Z}/(2)) \to 0$$

with the generator corresponding to $\{a, b\}$ in $\mathbf{Z}^{\binom{S}{2}}$ going to the element in $\mathbf{Z}^S$ given by the characteristic function of $\{a, b\}$. The map to $\mathbf{Z}^{\binom{S}{2}}$ gives the valuations along the divisors $D_{a,b}$ of Q^+.

3.15 Let $\chi_{s,t}$ be the character of $\mathbf{G}_m^S/\{\pm 1\}$ given by the characteristic function of $\{s, t\}$. The corresponding line bundle $\mathcal{L}$ on Q^+ is characterized by the existence of a $\chi_{s,t}$ equivariant map from $H^+/SL(2)$ to the total space of $\mathcal{L}$; one may take $\mathcal{L} = \mathcal{O}(-D_{s,t})$ and the map $F_{s,t}$. The isomorphism (3.14.2) of $Pic(Q^+)$ with $ker(\mathbf{Z}^S \to \mathbf{Z}/(2))$ is hence the opposite of the isomorphism 3.2.

Let $\mathbf{n} \in \mathbf{Z}^{\binom{S}{2}}$ be given. The conditions $\sum_t n_{s,t} = 0$ for the existence of $f \in \mathcal{O}^*(Q)$ with valuations $n_{s,t}$ along the $D_{s,t}$ can be interpreted as

$$\Sigma n_{s,t} D_{s,t} = 0 \text{ in } Pic(Q^+).$$

On H, there is up to a constant a unique function with valuation $n_{s,t}$ along the divisor $h_s = h_t$ of H^+. it is

$$F := \prod F_{s,t}^{n_{s,t}} \qquad \text{(product over } \binom{S}{2}\text{)}.$$

The conditions $\sum_t n_{s,t} = 0$ can also be interpreted as meaning that F is invariant by $\mathbf{C}^{*S}$, hence comes from Q.

§4. LAURICELLA'S HYPERGEOMETRIC FUNCTIONS

DEFINITION 4.1. *A local system of holomorphic functions over an analytic variety X is a* C-*linear subsheaf V of $\mathcal{O}$ which is a local system of finite dimensional* C-*vector spaces.*

If X is connected, a local section of V near $x_0 \in X$ extends as a multivalued holomorphic function on X. Conversely, let F be a multivalued holomorphic function, with the property that its branches at x_0 span a finite dimensional complex vector space. By analytic continuation, the same holds everywhere, and the functions on open subsets of X which, locally, are constant coefficient linear combinations of branches of F, form a local system of holomorphic functions.

4.2 If L is a rank one local system of holomorphic functions, the products ℓv, for ℓ (resp. v) a local section of L (resp. V) is again a local system of holomorphic functions, called the *twist* of V by L. If ℓ is a non zero multivalued section of L, it is also called the twist of V by ℓ.

Example Let f_k be non-vanishing holomorphic functions on X and $\alpha_k \in$ C. The determinations of $f := \prod_k f_k^{\alpha_k}$ are constant multiples of one another and hence span a rank one local system. The twist of V by this local system is denoted fV.

4.3 We fix a set S with $N \geq 4$ elements, a family μ of complex numbers indexed by S, with $\Sigma \mu_s = 2$ and $a, b, c \in S$. We will use the notations $S_1, M_1, Q, \dots$ of §1N.

We assume that no μ_s is an integer. For the integral case, see 4.13.

When x varies in M_1, the integrals from one ramification point x_s to another of a determination of

$$\omega := \prod_{s \neq c} (z - x_s)^{-\mu_s} \cdot dz$$

span a local system of holomorphic functions on M_1. The corresponding local system on Q is *the local system of Lauricella's hypergeometric functions*, relative to the choices of μ and of a, b, c.

4.4 We put together from [DM] 9.5, 4.6 the following results:

(a) The local system of Lauricella's hypergeometric functions is of rank $N - 2$. If F_i $(0 \leq i \leq N - 3)$ are linearly independent local sections, the F_i are the projective coordinates of an etale map from Q to $\mathbf{P}^{N-3}$ (cf. 6.3 below).

(b) Fix $s \in S_1$ and $t \in S, t \neq c$. Assume that $\mu_s + \mu_t \notin \mathbf{Z}$. Fix $q \in D_{s,t} \subset Q^+$ and a small ball B around q. On $B \cap Q$, one can find a multivalued basis F_i $(0 \leq i \leq N - 3)$ of the Lauricella's local system, i.e., a basis on the universal covering of $B \cap Q$, with the following properties.

(α) $F_1, \ldots, F_{N-3}$ are univalued and extend holomorphically across $D_{s,t}$;

(β) For z a local equation of $D_{s,t}$, F_0 is of the form $z^{1-\mu_s-\mu_t}$ (holomorphic invertible on B).

(γ) Let $\pi : S \to S_{st}$ be the quotient of S obtained by identifying s and t and define $\bar{\mu}$ on S_{st} by $\bar{\mu}_i = \sum_{\pi(j)=i} \mu_j : \bar{\mu}_{\pi(a)} = \mu_a$ for $a \neq s, t$ and $\bar{\mu}_{\pi(s)} = \mu_s + \mu_t$. We identify $D_{s,t}$ with $Q(S_{st})$. On $D_{s,t} = Q(S_{st})$, the F_i $(1 \leq i \leq N - 3)$ span the Lauricella hypergeometric local system, relative to $\pi(a), \pi(b), \pi(c)$ and $\bar{\mu}$.

4.5 We keep identifying Q with M_1 as in 4.3. Fix $a', b', c' \in S$ and let φ be the projectivity mapping $(x_{a'}, x_{b'}, x_{c'})$ to $(0, 1, \infty)$. It is the cross-ratio map

$$\varphi(z) = \frac{z - x_{a'}}{z - x_{c'}} : \frac{x_{b'} - x_{a'}}{x_{b'} - x_{x'}}.$$

The Lauricella's local system relative to μ and a', b', c' is obtained by integrating

$$\prod_{s \neq c'} (\varphi(z) - \varphi(x_s))^{-\mu_s} d\varphi(z).$$

This multivalued form still has valuation $-\mu_s$ on $\mathbf{P}^1$ at $z = x_s$; it is the product of $\prod_{s \neq c} (z - x_s)^{-\mu_s} dz$ by a product of functions $x_s, x_s - 1$ $(s \in S_1)$, $x_s - x_t$ $(s, t \in S_1)$ raised to complex powers depending on μ. One concludes

PROPOSITION 4.6. *For fixed μ, the Lauricella's local system relative to*

a', b', c' is the twist of the Lauricella local system relative to a, b, c by a multivalued function of the form $\prod_k f_k^{\alpha_k}$ with $f_k \in \mathcal{O}^*(Q)$ and $\alpha_k \in \mathbf{C}$.

The following definition is hence independent of the choice of a, b, c:

DEFINITION 4.7. *A hypergeometric-like local system relative to* μ, *is a twist of a Lauricella's local system relative to* μ *and* a, b, c *by* $\prod f_i^{\alpha_i}$ *with* $f_i \in \mathcal{O}^*(Q)$ *and* $\alpha_i \in \mathbf{C}$.

From 4.4 we now conclude, (as a, b, c can be freely chosen),

4.8 If V is hypergeometric-like relative to μ then

(a) A local basis $F_0 \ldots F_{N-3}$ of V defines an etale map from Q to $\mathbf{P}^{N-3}$ (cf. 6.3).

(b) Fix s, t distinct in S and assume that $\mu_s + \mu_t \notin \mathbf{Z}$.
Let $z \in \mathcal{O}^*(Q)$ have valuation 1 along $D_{s,t}$ (it can be defined by a cross ratio on M^+ for example); it is an equation for $D_{s,t}$ in $Q \cup D_{s,t} \subset Q^+$. As in 4.4 (b), one can find a multivalued basis F_i $(0 \leq i \leq N-3)$ of V near $q \in D_{s,t}$ such that, for suitable $\alpha_{s,t}$ and $\beta_{s,t}$ with $\beta_{s,t} - \alpha_{s,t} = 1 - \mu_s - \mu_t$ one has

(α) F_i $(1 \leq i \leq N-3)$ is of the form $z^{\alpha_{s,t}} f_i$ with f_i extending holomorphically across $D_{s,t}$.

(β) F_0 is of the form $z^{\beta_{s,t}} f_0$, with f_0 extending as an invertible holomorphic function across $D_{s,t}$.

(γ) On $D_{s,t}$, the f_i $(1 \leq i \leq N-3)$ span a hypergeometric-like local system relative to $\bar{\mu}$ (cf. 4.4 (γ)).

The exponents α, β are uniquely determined by V and μ: even if $N = 4$, in which case $(\alpha_{s,t}, \beta_{s,t})$ cannot be distinguished from $(\beta_{s,t}, \alpha_{s,t})$ in the behavior of V near $D_{s,t}$, the condition $\beta_{s,t} - \alpha_{s,t} = 1 - \mu_s - \mu_t$ tells which is α and which is β.

For Lauricella's local system relative to a, b, c and μ, one has by 4.4 (b) $\alpha_{s,t} = 0$ if $s \neq a, b, c$ and $t \neq c$; i.e., $\alpha_{s,t} = 0$ if $\{s, t\}$ is not $\{a, b\}$ or $\{u, c\}$ for some $u \in S$, $u \neq c$.

4.9 Making 4.5, 4.6 explicit, one could determine the $\alpha_{s,t}$ for the Lauricella's local system relative to a, b, c and then, using 3.5, obtain the possible $\alpha_{s,t}$ for a hypergeometric-like local system relative to μ.

Such a computation would lead to the answer: the only constraint on the $\alpha_{s,t}$ is that, for each $s \in S$

$$(4.9.1) \qquad\qquad \sum_t \alpha_{s,t} = \mu_s.$$

Let us admit (4.9.1). A more natural proof will be obtained in 5.8, using ideas of Gelfand et al and, in 4.12, we will sketch a proof in the spirit of what has been done so far.

If we twist a hypergeometric-like local system V relative to μ by $f = \prod f_i^{\alpha_i}$, having valuation $v_{s,t}$ along the $D_{s,t}$, the exponents $\alpha_{s,t}$ of (fV) are given in terms of those of V by

$$\alpha_{s,t}(fV) = \alpha_{s,t}(V) + v_{s,t}.$$

By the uniqueness assertion in 3.5, the twist fV of V is uniquely determined by its $\alpha_{s,t}$.

By (4.9.1), the μ_s are determined by the $\alpha_{s,t}$. This allows for the following definition.

DEFINITION 4.10. *Fix numbers* $\alpha_{s,t}$, *indexed by* $\{s,t\} \in \binom{S}{2}$, *with sum 1. Define* $\mu_s := \sum_t \alpha_{s,t}$; *their sum is 2. The* hypergeometric-like local system relative to α *is the hypergeometric-like local system relative to* μ *(4.7) whose exponents* $\alpha_{s,t}$ *are the given* $\alpha_{s,t}$; *we denote it* $L(\alpha)$.

4.11 If σ is in the symmetric group $\Sigma(S)$, σ acts on Q and, by transport of structure, carries the hypergeometric-like local system relative to σ into the hypergeometric-like local system relative to $\sigma(\alpha)$, with $\sigma(\alpha)$ defined by $\sigma(\alpha)_{\sigma(s),\sigma(t)} = \alpha_{s,t}$. This yields the following application.

Let H be a subgroup of the symmetric group $\Sigma(S)$. If α is H-invariant, the corresponding hypergeometric-like local system is stable by the action of H on Q.

If $N \geq 5$, H acts freely on a non-empty Zariski open subset Q' of Q. For $N = 4$, the Vierergruppe $V_4 \subset \Sigma(4)$ acts trivially on Q and we let Q' be the non-empty Zariski open subset of Q on which the quotient $H/H \cap V_4$ (which acts effectively on Q) acts freely. The local system V, being H-stable, is, restricted to Q', the inverse image of a local system of holomorphic functions on Q'/H; we will denote it V/H.

In the applications, we will also use that for any H-invariant system of numbers μ_s with sum 2, the equations (4.9.1), being affine linear, admit an H-invariant solution α: it can be obtained by averaging any solution over H.

4.12. Sketch of Proof of (4.9.1)

Let S_+ be deduced from S by the adjunction of an element w. Let $Q^+(S_+)$ and $Q^+(S)$ be the corresponding spaces Q^+. We have a smooth map

$$Q^+(S_+) \to Q^+(S).$$

In the definition 4.3 of the Lauricella's local system V relative to a, b, c and μ, we identify Q with M_1 and consider the fiber space with base M_1 and total space $\mathbf{P} \times M_1$ minus the sections x_s ($s \in S$); the functions in V are obtained by integration on chains in the fibers of a suitable multivalued relative 1-form ω. The fiber space can be viewed as being

$$Q(S_+) \to Q(S)$$

(identify $Q(S_+)$ with $M_1(S_+)$). Locally on $Q^+(S_+)$, the form ω is the product of a non-vanishing section of $\Omega^1_{Q^+(S_+)}/Q^+(S)$ by a product $\prod_k f_k^{\mu_k}$ with $f_k \in \mathcal{O}^*(Q(S_+))$ and $\mu_k \in \mathbf{C}$. It has a valuation $v_{s,t}(\omega)$ along each $D_{s,t}(s,t \in S_+)$.

Along $D_{s,w}$ the valuation of the form ω of 4.3 is $-\mu_s$. Along $D_{s,t}$ with $s, t \in S, s \in S_1, t \neq c$, the valuation is 0; it coincides with the exponent $\alpha_{s,t}$ of V (4.8).

Let f be a multivalued function on Q of the form $\prod_k f_k^{\alpha_k}$ with $f_k \in \mathcal{O}^*(Q)$ and $\alpha_k \in \mathbf{C}$. The twisted local system fV is obtained by integrating the multivalued form $\eta = f\omega$ on chains in the fibers. For $s, t \in S, s \in S_1, t \neq c$ the exponent $\alpha_{s,t}(fV) = v_{s,t}(f) + \alpha_{s,t}(V) = v_{s,t}(f)$ is equal to the valuation $v_{s,t}(f\omega) = v_{s,t}(f)$ of $f\omega$ along $D_{s,t} \subset Q^+(S_+)$.

If a, b, c are changed, the role of ω is played by another relative multivalued 1-form ω', and $f\omega = f'\omega'$, the multiple of ω' (4.5) whose integration yields fV. That is, independently of the choice of a, b, c, a hypergeometric-like local system V relative to μ (4.7) is obtained by integrating a multivalued form η on the chains in the fibers. For each choice of a, b, c, η can be written as $f\omega$ and

$$\alpha_{s,t}(V) = v_{s,t}(\eta),$$

for $s,t \in S, s \in S_1, t \neq c$. As a,b,c are freely chosen, the equality persists for all s,t. One also has

$$v_{s,w}(\eta) = -\mu_s.$$

Let ∇ be the unique connection on $\Omega^1_{Q(S_+)/Q(S)}$ for which, locally, the determinations of the multivalued relative 1-form η are horizontal. Such a connection exists because the determinations of η do not vanish and any two differ by a multiplicative constant. The line bundle $\Omega^1_{Q(S_+)/Q(S)}$ is the restriction to $Q(S_+)$ of $\Omega^1_{Q^+(S_+)/Q^+(S)}$ and the connection ∇ has a lograithmic pole along $D_{s,t}$, with residue (as defined in 2.9) given by $\mathrm{Res}(\nabla) = -v_{s,t}(\eta)$. Proposition 2.10 gives

$$c_1(\Omega^1_{Q^+(S_+/Q^+(S))}) = \Sigma v_{s,t}(\eta) c\ell(D_{s,t})$$

in $H^2(Q^+(S_+), \mathbf{C})$ as well as in $\mathrm{Pic}(Q^+(S_+)) \otimes \mathbf{C}$ by 2.6. Applying 3.11, we get

$$\sum_{t \in S_+} v_{s,t}(\eta) = \begin{cases} 0 & \text{if} \quad s \neq w \\ -2 & \text{if} \quad s = w. \end{cases}$$

This gives

$$\sum_{t \in S} \alpha_{s,t} - \mu_s = 0$$

(i.e., 4.9.1) and

$$\sum_{t \in S} -\mu_t = -2.$$

4.13 Following [DM] 2.15, we now explain how to modify the definitions 4.3 and 4.10 when some μ_s are integers. This will not really be needed in our main results, but it does allow for neater statements. We will be brief and we do not discuss here what happens along $D_{s,t}$ when $1 - \mu_s - \mu_t \in \mathbf{Z}$. We identify Q with M_1. On $\mathbf{P}^1 \times M_1/M_1$, we have sections x_s $(s \in S)$. Let L be the local system on the complement of these sections such that

(a) L is trivialized on the subset

$$\{(z,m) \mid z \text{ real}, \sup_{s \neq c} |x_s| < z < \infty\}$$

by a trivialization e;

(b) on each fiber $\mathbf{P}^1 - \{x_s(m) \mid s \in S\}$, L has monodromy $\alpha_s = \exp(2\pi i \mu_s)$ around $x_s(m)$.

The product

$$\omega := \prod_{s \neq c} (z - x_s)^{-\mu_s} \cdot e \cdot dz$$

extends as a L-valued relative 1-form. If μ_s is an integer, L extends across the section x_s; if further $\mu_s \leq 0$, ω extends; as a relative 1-form over M_1, its restriction to $x_s(M_1)$ is zero. Let $S' \subset S$ contain all s with μ_s an integer ≤ 0, and no s with μ_s an integer > 0. Let P' be the fiber space $P \times M_1$ over M_1 minus the sections x_s, $s \notin S'$; let $\pi' : P' \to M_1$ and let L' be the extension by 0 of L to P'. The relative 1-form ω has a class $[\omega]$ which is a section over M_1 of

$$(R^1 \pi'_* L') \otimes_{\mathbf{C}} \mathcal{O}_{M_1}.$$

Put $S'' = S - S'$; define $\pi'' : P'' \to M_1$ and L'' by exchanging the roles of S' and S''. The dual of the local system $R^1 \pi'_* L'$ on M_1 is the local system $R^1 \pi''_* (L^\vee)''$, for $L^\vee$ the local system dual to L. One can view $[\omega]$ as a map

$$[\omega] : R^1 \pi''_* (L^\vee)'' \to \mathcal{O}_{M_1}.$$

The image of this map is Lauricella's local system of holomorphic functions on M_1. It does not depend on the choice of S' as above.

In more concrete terms, it is the following local system of integrals. One takes the $\mathbf{C}$-linear span of the

(a) integral from x_s to x_t, when neither μ_s nor μ_t is an integer > 0.

(b) residue at x_s, when μ_s is an integer > 0.

The properties 4.4 (a)(b) continue to hold. The local system obtained varies holomorphically with μ. If the decomposition $S = S' \amalg S''$ is kept fixed, this is clear, and if a decomposition $S' \amalg S''$ is allowable for one μ, it remains allowable for nearby μ.

Example 4.14 Take $\mu_s = 0$ for $s \in S - \{c\}$, $\mu_c = 2$. The corresponding local system is spanned by the functions 1 and z_s ($s \in S_1$).

Example 4.15 Let us explain the continuity in μ_s when μ_s becomes an integer > 0. The integral from x_t to x_s is first reinterpreted as an integral along a cycle Γ with coefficients in the dual local system $L^\vee$, as follows (cf. [DM], 2.6)

f a section of $L^{\vee}$ near y. The monodromy around x_s is responsible for the factor α_s^{-1}. The condition that $\Gamma := \Gamma' + \lambda\Gamma''$ be a cycle is: $1 = \lambda - \lambda\alpha_s^{-1}$. We let f vary holomorphically with μ. The cycle $(1 - \alpha_s^{-1})\Gamma$ tends to the circuit Γ'' (with constant coefficients around x_s) when $\alpha_s \to 1$, and the corresponding integral tends to a residue.

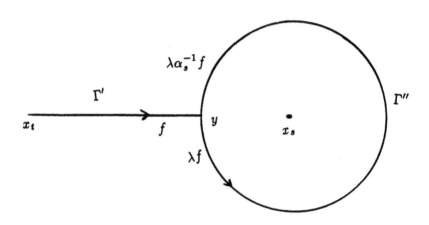

4.16 The monodromy of a hypergeometric-like local system on Q is a linear representation of $\pi_1(Q)$. The corresponding projective representations are the ones considered in [DM], for the same μ.

§5. GELFAND'S DESCRIPTION OF
HYPERGEOMETRIC FUNCTIONS

In their theory, Gelfand et al. consider local systems of holomorphic functions on a suitable Zariski open subset of
$Hom(\mathbf{C}^p, \mathbf{C}^N)$. For $p = 2$, it is an avatar of Lauricella's, and we will use
the pleasingly simple expression they get for the differential equations which
are satisfied.

5.1. We prefer using $\mathbf{C}^S$ to $\mathbf{C}^N$. With this change in notations, for
$p = 2$, the local system they consider is on the Zariski open subset $H \subset$
$\bar{H} = Hom(\mathbf{C}^2, \mathbf{C}^S)$ (3.12); it depends on the choice of complex numbers
$\mu_s (s \in S)$ with $\Sigma \mu_s = p = 2$. We assume at first that no μ_s is an integer.
For the general case, see 5.5.

We continue the notations of (3.12). On $\mathbf{C}^{*S}$, one fixes the multivalued
function
$z^{-\mu} := \Pi z_s^{-\mu_s}$. Its pullback by $h \in H$ is the multivalued function

$$h^{-\mu} := \Pi h_s^{-\mu_s}$$

on $\mathbf{C}^2$ minus the lines $h_s = 0$.

Let E be Euler's vector field $z_1 \partial_1 + z_2 \partial_2$ on $\mathbf{C}^2$. The 1-form

$$-E \perp (dz_1 \wedge dz_2) = z_2 dz_1 - z_1 dz_2$$

vanishes on each homogeneous line in $\mathbf{C}^2$ and is homogeneous of degree 2.
The product
$$\tilde{\eta}(h) := -h^{-\mu} E \perp (dz_1 \wedge dz_2)$$

is hence locally the pullback of a 1-form $\eta(h)$ on $\mathbf{P}^1$. Let h vary in a
small ball U around h_0. Let $(\mathbf{P}_U^1)'$ be the fiber space over U: complement
in $\mathbf{P}^1 \times U$ of the sections $h \to h_s^\perp (s \in S)$. The $\eta(h)$ define on $(\mathbf{P}_U^1)'$ a
multivalued relative 1-form. Taking suitable integrals of $\eta(h)$ - for instance

from one ramification point $h_s^{\perp}$ to another, as in 4.3 - one gets functions on U. These integrals span a local system of holomorphic functions on H, which we will denote $G(\mu)$.

5.2. We quote from [Gelfand, I.M. and Gelfand, S.I., 1986] the following differential equations satisfied by the local sections F of $G(\mu)$. We use the coordinates $h_{s,i}(s \in S, i = 1, 2)$ of 3.12 on H and we write $\partial_{s,i}$ for $\partial/\partial h_{s,i}$.

(5.2.1) $s\ell(2)$-invariance: F is fixed by the Lie algebra $s\ell(2)$.

(5.2.2)   homogeneity $-\mu_s$ in h_s:

$$(h_{s,1}\partial_{s,1} + h_{s,2}\partial_{s,2} + \mu_s)F = 0.$$

(5.2.3)   for s, t distinct in S, F is annihilated by the differential operator

$$\Delta_{s,t} := det \begin{pmatrix} \partial_{s,1} & \partial_{s,2} \\ \partial_{t,1} & \partial_{t,2} \end{pmatrix} = \partial_{s,1}\partial_{t,2} - \partial_{s,2}\partial_{t,1}.$$

The first two are obvious, with (5.2.1) resulting from the $SL(2)$-invariance of E and of $dz_1 \wedge dz_2$, hence of $E \perp (dz_1 \wedge dz_2)$. The third is familiar in Radon transform theory. In that theory, one fixes a C^{∞} function with fast decay ϕ on $\mathbf{R}^S$. The space H is replaced by $H_{\mathbf{R}} := Hom(\mathbf{R}^2, \mathbf{R}^S)$ and one defines on $H_{\mathbf{R}}$

(5.2.4) $$F(h) := \int_{\mathbf{R}^2} \phi(h(x_1, x_2))dx_1 \wedge dx_2.$$

This function $F(h)$ satisfies $\Delta_{s,t}F = 0$: one has

$$\partial_{t,2}F = \int_{\mathbf{R}^2} \partial_{t,2}\phi(h(x_1, x_2))dx_1 \wedge dx_2$$

$$= \int_{\mathbf{R}^2} (\partial_t\phi)(h(x_1, x_2)) \cdot x_2 \cdot dx_1 \wedge dx_2$$

$$\partial_{s,1}\partial_{t,2}F = \int_{\mathbf{R}^2} x_1 x_2 (\partial_s \partial_t \phi)(h(x_1, x_2))dx_1 \wedge dx_2.$$

The expression is symmetric in s and t and $\Delta_{s,t}F = 0$ follows. This elementary computation can be mimicked here as follows:

(a) On $(\mathbf{P}_U^1)'$ as in 5.1, let L be the local system of constant multiples of $\eta(h)$. The integral of $\eta(h)$ to be taken is an integral on a cycle representing an homology class with coefficients in L (with compact support). Such a cycle is given, in any fibre of $(\mathbf{P}_U^1)' \to U$, by oriented segments ℓ_i, avoiding the $h_s^\perp$, and constant multiples η_i of $\eta(h)$ on ℓ_i. A cycle condition as in 4.14 has to be met. The integral is $\Sigma \int_{\ell_i} \eta_i$. Two homologous cycles have the same integral.

(b) The cycle $C = \Sigma(\ell_i, \eta_i)$ on the fiber of $(\mathbf{P}_U^1)' \to U$ at h can be lifted to a cycle $\tilde{C}$ representing a homology class on $\mathbf{C}^2$ minus the $h_s = 0$, with coefficients in the local system which is the inverse image of L. This pullback of L is simply the local system $\tilde{L}$ of constant multiples of determinations of $\Pi h_s^{-\mu_s}$. The integral can be rewritten
$\Sigma \int_{\tilde{\ell}_i} h_i \cdot (-E \perp (dz_1 \wedge dz_2))$ where $\tilde{C} = \Sigma(\tilde{\ell}_i, h_i)$.

(c) If we multiply $\tilde{\ell}_i$ by $\lambda \in S^1 = \{z; |z| = 1\}$, it sweeps out a 2-chain $S^1 \tilde{\ell}_i$; taking on it a constant multiple h_i of a determination of $\Pi h_s^{-\mu_s}$, we get $\Sigma(S^1 \tilde{\ell}_i, h_i)$: a 2-cycle with coefficients in $\tilde{L}$. The integral on C can be rewritten using

$$\int_{S^1 \tilde{\ell}_i} h_i dz_1 \wedge dz_2 = \pm \int_{\tilde{\ell}_i} h_i (E \perp (dz_1 \wedge dz_2)) \int_{S^1} \frac{dz_1}{z_1} :$$

$$\Sigma \int_{\ell_i} \eta_i = \frac{\pm 1}{2\pi i} \Sigma \int_{S^1 \tilde{\ell}_i} h_i dz_1 \wedge dz_2$$

(sign depending on orientation conventions). This makes the integral formally similar to (5.2.4): it is of the form

$$\int_{cycle} z^{-\mu} \circ h \cdot dz_1 \wedge dz_2,$$

and the argument used in the real case works.

5.3. Fix elements a, b, c in S. We define $S_1 = S - \{a, b, c\}$ and let H_1 be the closed subset of H consisting of the h with

$$h_a = (1, 0) \qquad h_b = (1, -1) \qquad h_c = (0, 1)$$

and each $h_s (s \in S_1)$ of the form $(1, -x_s)$. On the projective line $\mathbf{P}^1 = (\mathbf{C}^2 - 0)/\mathbf{C}^*$ with coordinate z_1/z_2, one has $h_a^\perp = 0, h_b^\perp = 1, h_c^\perp = \infty$ and $h_s^\perp = x_s$ for $s \in S_1$. The action of $SL(2) \times \mathbf{C}^{*S}/(1,1), (-1,-1)$ (notation as in 3.12) on $\bar{H}$ induces an isomorphism

(5.3.1) $\qquad (SL(2) \times \mathbf{C}^{*S}/(1,1), (-1,-1)] \times H_1 \xrightarrow{\sim} H.$

By the properties (5.2.1)(5.2.2), the local system of holomorphic functions $G_1(\mu)$ is determined by its restriction to H_1.

For h in H_1, the bijection

(5.3.2)   (affine line $z_2 = 1$) - (points with $z_1 = 0, 1$ or a x_s)$\xrightarrow{\sim}$

$$\mathbf{P}^1 - (\text{points } h_s^\perp)$$

allows us to replace an integral in a cycle on $\mathbf{P}^1$ by an integral on $z_2 = 1$. The integrand - with pullback $-h^\mu \cdot E \perp (dz_1 \wedge dz_2)$ on $\mathbf{C}^2 - 0$, taken on the line $z_2 = 1$ with coordinate $z := z_1$ is

(5.3.3)
$$z^{-\mu_a}(z-1)^{-\mu_b} \prod_{s \in S}(z - x_s)^{-\mu_s} dx$$

The projection from H to $Q : h \to PGL(2)$-orbit of the system of points $h_s^\perp$ on $\mathbf{P}^1(s \in S)$ induces an isomorphism from H_1 to Q. By this isomorphism, the restriction $G_1(\mu)$ of $G(\mu)$ to H_1 becomes the local system of Lauricella's hypergeometric functions considered in 4.3.

COROLLARY 5.4. *The local system $G(\mu)$ on H is the pullback of the local system of Lauricella hypergeometric functions, twisted by* $\prod\limits_{s,t \in \binom{S}{2}} F_{s,t}^{-\alpha_{s,t}}$,

the exponents α being given by

$$\alpha_{a,b} = \frac{1}{2}(+\mu_a + \mu_b - \mu_c + \Sigma\mu_s)$$

$$\alpha_{a,c} = \frac{1}{2}(+\mu_a - \mu_b + \mu_c - \Sigma\mu_s)$$

$$\alpha_{b,c} = \frac{1}{2}(-\mu_a + \mu_b + \mu_c - \Sigma\mu_s) \text{ (sums on } S_1)$$

$$\alpha_{c,s} = \mu_s, (s \in S_1)$$

$$\alpha_{s,t} = 0 \text{ otherwise.}$$

Remark The values of α given above are obtained by solving the equations

$$\Sigma_t \alpha_{s,t} = \mu_s, \text{ for } s \in S, \text{ and}$$

$$\alpha_{s,t} = 0 \text{ for } s \in S_1 \text{ and } t \neq c$$

i.e., $\{s,t\} \neq \{a,b\}, \{a,c\}, \{b,c\}$ or $\{c,s\}$ for some $s \in S_1$.

PROOF:. The factors $F_{a,b}, F_{a,c}, F_{b,c}, F_{c,s}$ are ± 1 on H_1. Their homogeneity on H is known and they are $SL(2)$-invariant. The pull-back of the Lauricella local system, multiplied by the given factor, is hence $SL(2)$-invariant and one checks it has homogeneity $-\mu_s$ in $h_s(s \in S)$, using $\Sigma_t \alpha_{s,t} = \mu_s$. By (5.3.1), (5.2.1) and (5.2.2), it coincides with $G(\mu)$.

5.5. We now allow $\mu_s \in \mathbf{Z}$. The integrals to be taken in the definition of $G(\mu)$ are to be modified as in 4.12. The relation with Lauricella's local system remains as in 5.3, 5.4 and $G(\mu)$ depends holomorphically on μ. By continuity, the properties 5.2 continue to hold.

COROLLARY 5.6. *Assume* $1 - \mu_s - \mu_t \notin \mathbf{Z}$. *Then, for B a small ball around a point q of the divisor $F_{s,t} = 0$ of H^+ (notation as in 3.12), the local system $G(\mu)$ has a multivalued basis $F_0, \ldots, F_{N-3}$ for which*

(a) $F_1, \ldots, F_{N-3}$ extend holomorphically across $F_{s,t} = 0$ and, on the divisor $F_{s,t} = 0$, are the homogeneous coordinates of a submersive map from the divisor $F_{s,t} = 0$ to $\mathbf{P}^{N-4}$.

(b) F_0 is of the form $F_{s,t}^{1-\mu_s-\mu_t} f_0$ with f_0 extending as an invertible holomorphic function across $F_{s,t} = 0$.

PROOF. We choose a, b, c such that $s \in S_1$ and $t \neq c$. Then apply 4.4 and 5.4.

COROLLARY 5.7. *Let V be the Lauricella local system on Q relative to μ and a, b, c. For all s, t in S, the exponent $\alpha_{s,t}$ of V along $D_{s,t}$, in the sense of 4.8, is given by the formula of 5.4.*

The exponent $\beta_{s,t}$ is then given by $\beta_{s,t} - \alpha_{s,t} = 1 - \mu_s - \mu_t$.

PROOF. Let $\alpha'_{s,t}, \beta'_{s,t} = \alpha'_{s,t} + (1 - \mu_s - \mu_t)$ be the exponents of V along $D_{s,t}$ as in 4.8. For q on the divisor $F_{s,t} = 0$ of H^+ and z a local equation for this divisor, the local system V, pulled back to H and twisted by $z^{-\alpha'_{s,t}}$ satisfies the properties (a), (b) of 5.6. If on the other hand, we twist by $\Pi F_{s,t}^{-\alpha_{s,t}}$, one obtains $G(\mu)$ which also has the properties 5.6. Since one is a twist of the other, we conclude that $z^{-\alpha'_{s,t}}$ and $\Pi F_{s,t}^{-\alpha_{s,t}}$ have the same valuation along $F_{s,t} = 0$, i.e., that $\alpha_{s,t} = \alpha'_{s,t}$.

COROLLARY 5.8. *Let V be a hypergeometric-like local system on Q, relative to μ. Then, its exponents 4.8 satisfy*

(5.8.1) $$\Sigma_t \alpha_{s,t} = \mu_s.$$

PROOF. The local system V is a twist of the Lauricella local system relative to μ and a, b, c. By 3.5, it suffices to check (5.8.1) for the latter, in which case it results from 5.7, indeed, the exponents $\alpha_{s,t}$ of 5.4 were constructed to satisfy (5.8.1).

PROPOSITION 5.9. *The differential equations (5.2.1), (5.2.2) and (5.2.3) are enough to express each second derivative of F as a linear combination of F and its first derivatives.*

PROOF. We first prove it on the open subset U of H where each $h_{s,i} \neq 0$. We will write ... for a linear combination with holomorphic (even algebraic) coefficients of F and the $\partial_{s,i}F$. Applying $\partial_{t,i}$ to (5.2.2), we get

$$\partial_{t,1}\partial_{s,2}F = -h_{s,1}/h_{s,2}\partial_{t,1}\partial_{s,1}F + \ldots$$
$$\partial_{t,2}\partial_{s,2}F = -h_{s,1}/h_{s,2}\partial_{t,2}\partial_{s,1}F + \ldots = -h_{s,1}/h_{s,2}\partial_{s,1}\partial_{t,2}F + \ldots$$
$$= \frac{h_{s,1}h_{t,1}}{h_{s,2}h_{t,2}}\partial_{s,1}\partial_{t,1}F + \ldots$$

The equation (5.2.3) becomes

$$(-h_{t,1}/h_{t,2} + h_{s,1}/h_{s,2})\partial_{s,1}\partial_{t,1}F + \ldots = 0,$$

and the coefficient may be rewitten $F_{s,t}/h_{s,2}h_{t,2}$. Thus $\partial_{s,1}\partial_{t,1}F$ can be expressed in the desired form on U. Finally (5.2.1) gives in particular

$$\Sigma_t h_{t,2}\partial_{t,1}F = 0,$$

hence for any $s \in S$

$$h_{s,2}\partial_{s,1}^2 F = -\Sigma_{t \neq s} h_{t,2}\partial_{s,1}\partial_{t,1}F + \ldots$$

Thus all second derivatives may be expressed in the desired form on U. The system of equations (5.2) is $SL(2)$-invariant. For any $g \in SL(2, \mathbf{C})$, the conclusion of 5.3 holds therefore on gU. Since the gU cover H, the proposition follows.

Remark The precise meaning of 5.9 is: in the ring $\mathcal{D}_H$ of algebraic differential operators on the affine variety H, the left ideal I generated by

$$\Sigma_s h_{s,2}\partial_{s,1}, \Sigma_s h_{s,i}\partial_{s,2}, \Sigma_s h_{s,1}\partial_{s,1} - h_{s,2}\partial_{s,2},$$
$$h_{s,1}\partial_{s,1} + h_{s,2}\partial_{s,2} + \mu_s(s \in S), \text{ and } \Delta_{s,t}(s,t \in \binom{S}{2})$$

is such that

$$I + (\text{operators of order } \leq 1) = \mathcal{D}_H.$$

5.10 Fix complex number $\alpha_{s,t}(\{s,t\} \in \binom{S}{2})$ with $\Sigma_t \alpha_{s,t} = \mu_s$. The local system $G(\mu)$, twisted by $\prod_{s,t} F_{s,t}^{\alpha_{s,t}}$, is a local system of functions which are $SL(2)$-invariant and of homogeneity zero in each h_s. As $SL(2) \times \mathbf{C}^{*S}$ is connected, it is the pullback of a local system of holomorphic functions on Q. If the $\alpha_{s,t}$ are as in 5.4, 5.4 implies that one gets the Lauricella local system of hypergeometric functions relative to (a,b,c) and μ (4.3). Let α^0 be this particular choice of exponents. For another choice of α, with $\Sigma_t \alpha_{s,t} = \Sigma_t \alpha_{st}^0 = \mu_s$, the corresponding local system on Q is the twist of Lauricella's by the multivalued function f with pullback on H the product $\prod F_{s,t}^{\alpha_{s,t} - \alpha_{s,t}^0}$. This valuation of f along $D_{s,t}$ is $\alpha_{s,t} - \alpha_{s,t}^0$. This construction provides all hypergeometric-like local systems on Q relative to μ.

Our description of the relation between twisting on H and on Q shows that

$$\prod F_{s,t}^{\alpha_{s,t}} G(\mu)$$

is the pullback to H of the hypergeometric-like local system on Q relative to α defined in §4.

5.11. Fix complex numbers $\alpha_{s,t}, s,t \in \binom{S}{2}$, with sum 1. Define $\mu_s :=$ $\Sigma_t \alpha_{s,t}$. Take $q \in Q$, m above q in H. Choose a determination F of $\prod F_{a,b}^{-\alpha_{a,b}}$ near m. For f a holomorphic function on Q defined in a neighborhood of Q, consider on H, in a neighborhood of m

$$F^{-1} F_{s,t} \Delta_{s,t}(Ff);$$

here we write f for the pullback of f. This function is independent of the choice of F. It is $SL(2)$-invariant and of homogeneity 0 in each h_s. Hence it comes from a function

$$\Xi_{s,t}^{\alpha}(f)$$

on Q. The operator $\Xi_{s,t}^{\alpha}$ is a second order differential operator on Q. By 5.2.3 and 5.10, one has

PROPOSITION 5.12. *The functions f belonging to the local system $L(\alpha)$ of 4.10 satisfy the equations $\Xi^{\alpha}_{s,t}(f) = 0$ (for $s,t \in \binom{S}{2}$).*

As a result of 5.9 these equations are enough to express each second derivative of f in terms of f and its first derivatives.

§6. STRICT EXPONENTS

In this section, we describe singularities of local systems of holomorphic functions along a divisor, of the type encountered for Lauricella's hypergeometric functions.

6.1. Let V be a local system of holomorphic functions on a non-singular analytic variety X. We will always identify a vector bundle with its $\mathcal{O}$-module sheaf of holomorphic sections. Set $\mathcal{V} = \mathcal{O} \otimes_{\mathbb{C}} V$. The vector bundle $\mathcal{V}$ carries an integrable connection ∇ for which V is the local system $\mathcal{V}^{\nabla}$ of horizontal sections. Namely, $\nabla(f \otimes v) = df \otimes v$ for $v \in V$ and $f \in \mathcal{O}$.

The inclusion of V in $\mathcal{O}$ extends by $\mathcal{O}$-linearity to $\lambda_{\mathcal{V}}$ or simply $\lambda : \mathcal{V} \rightarrow \mathcal{O}$. From the vector bundle $\mathcal{V}$, the integrable connection ∇ and the linear form λ, one recovers $V \subset \mathcal{O}$ as being $\lambda(\mathcal{V}^{\nabla})$.

The bundle of first order jet of functions $\mathrm{jet}_1(\mathcal{O})$ can be identified with $\mathcal{O} \oplus \Omega^1$, with $f \mapsto \mathrm{jet}_1(f)$ being $f \mapsto (f, df)$. The $\mathbb{C}$-linear map $f \mapsto \mathrm{jet}_1(f)$ from V to $\mathrm{jet}_1(\mathcal{O})$ extends by $\mathcal{O}$-linearity to a map $\lambda_1 : \mathcal{V} \rightarrow \mathrm{jet}_1(\mathcal{O})$. Its components relative to the isomorphism of $\mathrm{jet}_1(\mathcal{O})$ with $\mathcal{O} \oplus \Omega^1$, are given by

$$(6.1.1) \qquad\qquad \lambda_1 = (\lambda, d\lambda - \lambda\nabla).$$

Example 6.2 Let L be a rank one local system of invertible holomorphic functions. This is an abuse of language for: local generators ℓ of L are nowhere vanishing. Let $\mathcal{L}$ be the corresponding vector bundle $\mathcal{O} \otimes_{\mathbb{C}} L$. The 1-form $\nu := \frac{d\ell}{\ell}$ for ℓ a local generator of L is independent of ℓ, hence globally defined, and L is the local system of the functions f for which $df = \nu f$. The map $\lambda_L : \mathcal{L} \rightarrow \mathcal{O}$ is an isomorphism and carries the connection ∇ of $\mathcal{L}$ into the connection $d - \nu$ of $\mathcal{O}$. We call ν *the 1-form defining L*.

Suppose now that $(D_i)_{i \in I}$ is a family of disjoint divisors on the analytic variety X, L is a local system of invertible holomorphic functions on $Y :=$ $X - \bigcup_i D_i$ and $\mathcal{L}$ is the corresponding vector bundle on Y.

Let $\mathcal{L}_X$ denote an extension of $\mathcal{L}$ to a line bundle on X. If λ_L extends to an isomorphism of $\mathcal{L}_X$ with $\mathcal{O}$, then the 1-form of 2.10.2 whose residue gives $-\text{Res}_{D_i}\nabla$ coincides with the 1-form defining L, since $\nabla = d - \nu$.

Fix q in the smooth locus of D_i and let z be a local equation for D_i near q. The following conditions are equivalent:

(a) ν has a simple pole along D_i near q

(b) near q, a multivalued section ℓ of L can be written $\ell = z^\alpha f$, where f extends as an holomorphic invertible function across D_i.

The exponent α is the *valuation* of ℓ along D_i (cf. 2.8); it is the residue of ν along D_i. If the condition (a, b) holds along all the D_i, one can rephrase (2.11) in the form: the valuations $v_i(\ell)$ along the D_i of a multivalued section ℓ of L satisfy

$$\Sigma \, v_i(\ell) \cdot cl(D_i) = 0$$

in $H^2(X, \mathbf{C})$ (cf. 2.7).

Consider $V, \mathcal{V}, \lambda, \lambda_1$ as in 6.1. Let V^L be the twist of V by L, and $\mathcal{V}^L, \lambda^L, \lambda^L$ be the corresponding vector bundle and linear maps. One has a canonical isomorphism

$$(6.2.1) \qquad \mathcal{V}^L = \mathcal{O} \otimes (V \otimes_{\mathbf{C}} L) = V \otimes \mathcal{L} \overset{\lambda^L}{\to} \mathcal{V};$$

it carries $f \otimes (\ell v)$ into $(f\ell) \otimes v$. If we identify $\mathcal{V}$ and $\mathcal{V}^L$, the connection ∇^L of $\mathcal{V}^L$ and the linear maps $\lambda^L, \lambda_1^L = (\lambda^L, d\lambda^L - \lambda^L\nabla^L)$ are given by

$$(6.2.2) \qquad \nabla^L(w) = \nabla w - \nu w$$

$$(6.2.3) \qquad \lambda^L = \lambda \qquad \lambda_1^L(w) = \lambda_1(w) + \lambda(w) \cdot \nu.$$

We now assume that X is purely of dimension n and that V is of dimension $n + 1$.

PROPOSITION 6.3. *The following conditions are equivalent:*

(i) $\lambda_1 : \mathcal{V} \to jet_1(\mathcal{O})$ *is an isomorphism*

(ii) for any local basis $e_0, \dots, e_n$ *of V, the functions* $e_0, \dots, e_n$ *are the projective coordinates of an etale map from X to $\mathbf{P}^n$.*

PROOF. In local coordinates, the matrix of λ_1 is the $(n+1) \times (n+1)$ matrix

$$(6.3.1) \qquad D\begin{pmatrix} e_0, & \ldots, e_n \\ 1, & \partial_1, \ldots, \partial_n \end{pmatrix} := \begin{pmatrix} e_0 & \cdots & e_n \\ \partial_1 e_0 & \cdots & \partial_1 e_n \\ \vdots & & \vdots \\ \partial_n e_0 & \cdots & \partial_n e_n \end{pmatrix}$$

Etaleness of $(e_0, \ldots, e_n)$ at $p \in X$ means that for no nonzero tangent vector v at p is the vector $(\partial_v e_0, \ldots, \partial_v e_n)$ proportional to the vector $(e_0, \ldots, e_n)$ at p. This amount to the linear independence of the rows of (6.3.1), hence 6.3.

COROLLARY 6.4. *If V^L is the twist of V by a local system of invertible holomorphic functions (4.2), and if V satisfies the equivalent conditions of 6.3, so does V^L.*

PROOF. The map in 6.3(ii) is the same for V and V^L.

Remark 6.5 Let $\lambda_k : V \to \mathrm{jet}_k(\mathcal{O})$ be the extension by $\mathcal{O}$-linearity of the map $f \to \mathrm{jet}_k(f) : V \to \mathrm{jet}_k(\mathcal{O})$. If 6.3 (i) holds, the map $\mu := \lambda_2 \lambda_1^{-1} :$ $\mathrm{jet}_1(\mathcal{O}) \to \mathrm{jet}_2(\mathcal{O})$ is defined. It is a section of the natural projection $\mathrm{jet}_2(\mathcal{O}) \to \mathrm{jet}_1(\mathcal{O})$ and for any function u in V, one has

$$(6.5.1) \qquad \mathrm{jet}_2(u) = \mu \, \mathrm{jet}_1(u).$$

The second derivatives of u are expressed as an $\mathcal{O}$-linear combination of u and its first derivatives. In fact, V is the local system of all solutions of the system of linear differential equations (6.5.1).

A section μ of $\mathrm{jet}_2(\mathcal{O}) \to \mathrm{jet}_1(\mathcal{O})$ defines a connection ∇_μ on $\mathrm{jet}_1(\mathcal{O})$, characterized by the property that for any f, $\mathrm{jet}_1(f)$ (a section of $\mathrm{jet}_1(\mathcal{O})$) satisfies $\nabla_\mu \mathrm{jet}_1(f) = 0$ at p if and only if $\mathrm{jet}_2(f) = \mu(\mathrm{jet}_1(f))$ at p. The horizontal sections of ∇_μ are the $\mathrm{jet}_1(f)$ for f a solution of (6.5.1) and, if ∇_μ is integrable, the solutions of (6.5.1) constitute a local system of holomorphic functions for which (6.3) (i) holds.

In local coordinates, the 1-jet of a function is defined by its value u and its first derivatives v_i, and (6.5.1) is a system of equations

$$\partial_i \partial_j f = A_{ij}^0 f + \Sigma \, A_{ij}^k \partial_k f$$

with $A_{ij}^a = A_{ji}^a$; the equation $\nabla_\mu = 0$ is the first order system

$$\begin{cases} \partial_i u = v_i \\ \partial_i v_j = A_{ij}^0 u + \Sigma A_{ij}^k v_k. \end{cases}$$

6.6 Let X be an analytic variety of dimension n, D be a smooth divisor on X and let V be a local system of holomorphic functions on $X - D$, of dimension $n + 1$. Fix two complex numbers α and β. We will assume that $\alpha - \beta \notin \mathbf{Z}$.

For $p \in D$ and z a local equation for D at p, we consider the condition:

6.6.1. There are holomorphic functions $g_i(0 \leq i \leq n)$, defined in a neighborhood U of p, such that locally on $U - D$, $e_0 = z^\beta g_0$ and the $e_i = z^\alpha g_i(1 \leq i \leq n)$ form a basis of V.

The condition is independent of the choice, made locally on $U - D$, of the determinations of z^β and z^α. It is also independent of the choice of z, the local equation. For U a small ball around p, the $e_i(0 \leq i \leq n)$ are a multivalued basis of V on $U - D$, i.e., a basis of the pullback of V on the universal covering of $U - D$.

For U a small ball around p, the fundamental group of $U - D$ is $\mathbf{Z}$; generator: a positive turn around D. Put $a = exp(2\pi i\alpha)$ and $b = exp(2\pi i\beta)$. Relative to the basis e_i, the local monodromy around D is the diagonal matrix with diagonal $(b, a, \ldots, a)$. On $U - D$, the local monodromy around D is an automorphism of V. It defines an eigenspace decomposition of $V : V = V' \oplus V''$ with V' spanned by e_0 and V'' spanned by $e_i(1 \leq i \leq n)$.

In a basis free language, (6.6.1) can be restated:

6.6.2. In a neighborhood of D, the local system V is a direct sum $V' \oplus V''$, with V' (resp. V'') of dimension 1 (resp. n) and with local monodromy around D the scalar b (resp. a). For z a local equation for D in a neighborhood $U(p)$ of $p \in D$, the local system $z^{-\beta}V'$ and $z^{-\alpha}V''$ extend across D as local systems of holomorphic functions $(z^{-\beta}V')_{U(p)}$ and $(z^{-\alpha}V'')_{U(p)}$ on $U(p)$.

If $\beta = 0$ (resp. $\alpha = 0$), $(V')_{U(p)}$ (resp. $(V'')_{U(p)}$) is independent of the choice of z. It then globalizes to a local system defined in a neighborhood of D.

We will sometimes supplement (6.6.1) by the condition

(6.6.3) Neither g_0 nor any non-trivial constant coefficient linear combination of the

$g_i(1 \leq i \leq n)$ vanishes everywhere on D.

In the basis free language of (6.6.2), (6.6.3) means: the restrictions to D
of the functions in
$(z^{-\beta}V')_{U(p)}$ (resp. $(z^{-\alpha}V'')_{U(p)}$) form a local system of rank 1 (resp. n).
If $\beta = 0$ (resp. $\alpha = 0$), this restriction globalizes to a local system V_D' (resp
V_D'') on D.

Assume (6.6.3). If $D \neq \phi$ and $n \geq 2$, the exponents α and β are
then uniquely determined by V. For $n = 1$, they are determined up to
permutation.

(6.7) We assume (6.6.1).

The vector bundle $V = \mathcal{O} \otimes_{\mathbb{C}} V$ on $X - D$ is extended to a vector bundle
V_X on X by taking the single valued basis $(z^{-\beta} \otimes e_0, z^{-\alpha} \otimes e_1, \ldots, z^{-\alpha} \otimes e_n)$
as a basis for the extension around $p \in D$ (notations as in 6.6). Let j denote
the inclusion of $X - D$ in X. In a neighborhood of D where one has a
decomposition $V = V' \oplus V''$ as in 6.6, a basis free description of V_X is: the
subsheaf of $\mathcal{O}$-modules of j_*V spanned by the $z^{-\beta} \otimes e$ and the $z^{-\alpha} \otimes f$ for
z a local equation of D, e a multivalued section of V' and f a multivalued
section of V''.

We define
$$\text{jet}_1(\mathcal{O})(\log\ D) := \mathcal{O} \oplus \Omega^1(\log\ D).$$

PROPOSITION 6.8. *With the notations of 6.7*
(i) The morphism $\lambda : V \to \mathcal{O}$ on $X - D$ extends as $\lambda : V_X \to \mathcal{O}$ on X.
(ii) The connection ∇ extends as $\nabla : V_X \to \Omega^1(\log\ D) \otimes V_X$.
(iii) The morphism λ_1 extends as $\lambda_1 : V_X \to \text{jet}_1(\mathcal{O})(\log\ D)$.

PROOF. (i) $\lambda(z^{-\beta} \otimes e_0)$ and the $\lambda(z^{-\alpha} \otimes e_1)(1 \leq i \leq n)$ are the holomorphic functions g_i

$$\nabla(z^{-\beta} \otimes e_0) = -\beta z^{-\beta-1} dz \otimes e_0$$

(ii)
$$= -\beta \frac{dz}{z} \otimes z^{-\beta} \otimes e_0 \text{ and similarly}$$

$$\nabla(z^{-\alpha} \otimes e_i) = -\alpha \frac{dz}{z} \otimes z^{-\alpha} \otimes e_i.$$

$$\lambda_1(z^{-\beta} \otimes e_0) = z^{-\beta}\mathrm{jet}_1(z^{\beta}g_0)$$

(iii)
$$= z^{-\beta} \cdot (z^{\beta}g_0, z^{\beta}dg_0 + \beta \, z^{\beta-1}dz \cdot g_0)$$

$$= (g_0, dg_0 + \beta \, \frac{dz}{z} \cdot g_0)$$

and similarly for the $z^{-\alpha} \otimes e_i (1 \le i \le n)$. One could also deduce (iii) from (i), (ii) and 6.2.1.

DEFINITION 6.9. *Let V on $X - D$ be such that $\lambda_1 : V \to jet_1(\mathcal{O})$ is an isomorphism. We say that V has* strict exponents (α, β) *along D if (6.6.1) holds and if, with the notations of 6.8,*

$$\lambda_1 : \mathcal{V}_X \to jet_1(\mathcal{O})(\log \, D)$$

is an isomorphism.

6.10. Let L be a rank one local system of invertible holomorphic functions on $X - D$. Assume that the closed 1-form ν defining L (6.2) has a simple pole along D. Let γ be its residue (a constant, at least if D is connected). If $U(p)$ is a small neighborhood of $p \in D$, this means that multivalued sections f of L on $U(p) - D$ are of the form $z^{\gamma}u$, with z a local equation for D and u holomorphic invertible.

If V satisfies (6.6.1) (resp. and (6.6.3)) along D, for the exponents (α, β), then the twist V^L of V by L satisfies the same conditions for the exponents $(\alpha + \gamma, \beta + \gamma)$. The isomorphism (6.2.1) between V and V^L extends to an isomorphism on X between $\mathcal{V}_X$ and $\mathcal{V}_X^L$. By (6.2.3), the map λ_1^L is obtained by composing λ_1 with the automorphism $(u, v) \to (u, v + \nu u)$ of $jet_1(\mathcal{O})(\log \, D)$. One concludes

PROPOSITION 6.11. *With notations in 6.10, if V has strict exponents (α, β) along D, then V^L has strict exponents $(\alpha + \gamma, \beta + \gamma)$ along D.*

PROPOSITION 6.12. *Assume (6.6.1). Then, with the notations of 6.6, V has strict exponents (α, β) along D if and only if*

(a) g_0 is invertible on D.

(b) the $g_i (1 \le i \le n)$ are the projective coordinates of an etale map from D to $\mathbf{P}^{n-1}$.

PROOF. Let $(z_1, \ldots, z_n)$ be a system of local coordinates with $z_1 = z$. With respect to the base $(z^{-\beta} \otimes e_0, z^{-\alpha} \otimes e_1, \ldots, z^{-\alpha} \otimes e_n)$ of $\mathcal{V}_X$ and $(1, \frac{dz}{z}, dz_2, \ldots, dz_n)$ of $\mathrm{jet}_1(\mathcal{O})(\log D)$, the columns of the matrix of λ_1 are (written horizontally)

$$z^{-\beta}(1, z_1 \partial_1, \partial_2, \ldots \partial_n) z^{\beta} g_0 \text{ and the}$$
$$z^{-\alpha}(1, z_1 \partial_1, \partial_2, \ldots, \partial_2) z^{\alpha} g_i \ (i \neq 0).$$

On D, this becomes

$$(1, \beta, \partial_2, \ldots, \partial_n) g_0 \text{ and the}$$
$$(1, \alpha, \partial_2, \ldots, \partial_n) g_i \ (i \neq 0).$$

A linear combination of the first two rows is $(g_0, 0, \ldots, 0)$. (We are assuming $\alpha - \beta \neq \mathbf{Z}$, in particular $\alpha \neq \beta$). One concludes that λ_1 is invertible on D if and only if g_0 and $D \begin{pmatrix} g_1 & & g_n \\ 1, & \partial_2, \ldots, & \partial_n \end{pmatrix}$ are. The invertibility of the latter means that $g_1, \ldots, g_n$ are the projective coordinates of an etale map from D to $\mathbf{P}^{n-1}$ (cf. 6.3).

COROLLARY 6.13. *If V has strict exponents (α, β) along D, the condition (6.6.3) holds.*

Remark 6.14 With the notation of 6.10, each g_i changes to ug_i when V is replaced by V^L, and 6.12 gives another proof of 6.11.

Remark 6.15 If $\beta = 0$, the condition (a) of 6.12 means that the local system V'_D on D (6.6) is a rank one local system of invertible holomorphic functions. If $\alpha = 0$, the condition (b) of 6.12 means that the rank n local system V''_D on D (6.7) satisfies the equivalent conditions of 6.3. Locally along D, such a vanishing can be achieved by replacing V by $z^{-\beta} V$ or by $z^{-\alpha} V$ respectively, for z a local equation of D.

Applying 4.8, we get our main example.

Example 6.16 With the notations of 4.7 or 5.9, and assuming $\mu_s + \mu_t \notin \mathbf{Z}$, hypergeometric-like local systems of holomorphic functions, rel. μ, on $Q \subset Q^+$ have strict exponents $(\alpha_{s,t}, \beta_{s,t})$ along the divisors $D_{s,t}$. One has $\beta_{s,t} - \alpha_{s,t} = 1 - \mu_s - \mu_t$.

6.17. We will need a variant of 6.9 for the case $\alpha = 0$, $\beta = 1$. Let V be a local system of holomorphic functions on $X - D$ and assume that

(a) λ_1 is an isomorphism (on $X - D$);

(b) V extends as V_X on X.

We also assume that

(c) near any $p \in D$, V_X has a local section e_0 which is a holomorphic function vanishing identically on D.

We complete e_0 to a local basis $(e_0, e_1, \ldots, e_n)$ and choose a local equation z for D. The following conditions are then equivalent

(6.17.1) For V_X on X, $\lambda_1 : \mathcal{O} \otimes_{\mathbf{C}} V_X \to \mathrm{jet}_1(\mathcal{O})$ is an isomorphism.

Let $\mathcal{V}_X$ be the extension of V across D with basis $(z^{-1} \otimes e_0, e_1, \ldots, e_n)$.

(6.17.2) $\lambda_1 : \mathcal{V}_X \to \mathrm{jet}_1(\mathcal{O})(\log D)$ is an isomorphism.

(6.17.3) On D, e_0/z is invertible and $(e_1, \ldots, e_n)$ are the projective coordinates of an etale map from D to $\mathbf{P}^{n-1}$.

PROOF. We choose local coordinates $(z_1, \ldots, z_n)$ with $z_1 = z$. Condition 6.17.1 is

$$D \begin{pmatrix} e_0 \ldots e_n \\ 1, \partial_1, \ldots \partial_n \end{pmatrix} \neq 0.$$

On D, e_0 and the $\partial_i e_0$ for $i \neq 1$ vanish, and $\partial_1 e_0 = e_0/z$. Invertibility hence means that e_0/z is invertible, and that

$$D \begin{pmatrix} e_1 \ldots e_n \\ 1, \partial_2, \ldots \partial_n \end{pmatrix} \neq 0.$$

The proof of 6.12 for $\alpha = 0$ and $\beta = 1$ shows that those conditions are equivalent to (6.17.2) and (6.17.3).

DEFINITION 6.18. *When the conditions (a) and (b) of 6.17 are satisfied, we say that V (or V_X) has strict exponents $(0, 1)$ along D if (c) and the equivalent conditions $(6.17.i)$, $i = 1, 2, 3$ hold.*

6.19. Let now $\tau : X \to Y$ be a map ramified along D, with ramification index r. The questions we consider are local along D, so that we may assume that τ is $(z_1, \ldots z_n) \mapsto (z_1^r, z_2, \ldots, z_n)$, with z_1 an equation for D. Let E be the image of D. Let V_X be a local system of holomorphic functions on $X - D$, which is the pullback of a local system of holomorphic functions V_Y on $Y - E$. We assume V_X of dimension $n + 1$ and that λ_1 is an isomorphism for V_Y and hence for V_X.

PROPOSITION 6.20. *(i)* V_X *extends to a local system of holomorphic functions on* X *for which* λ_1 *is an isomorphism if and only if* V_Y *has strict exponents* $(0, 1/r)$ *along* E.

(ii) V_X *has strict exponents* (α, β) *along* D *if and only if* V_Y *has strict exponents* $(\alpha/r, \beta/r)$ *along* E.

Note: In (ii) and elsewhere, we assume $\alpha - \beta \notin \mathbf{Z}$ for "strict exponent" to be defined, unless there is explicit mention to the contrary.

PROOF. Locally Y is the quotient of X by the cyclic group with generator

$T : (z_1, \dots, z_n) \mapsto (\zeta z_1, z_2, \dots, z_n)$, $\zeta = \exp(2\pi i/r)$. On the fixed point set D, T acts on $\mathrm{jet}_1(\mathcal{O})$ as a complex reflection with eigenvalue ζ^{-1}. Here and later, the action is by transport of structure: on functions, T acts by $f \mapsto f \circ T^{-1}$; hence the inverse ζ^{-1}.

Let us first assume that V_X extends to a local system of holomorphic functions V on X and that $\lambda_1 : V \to \mathrm{jet}_1(\mathcal{O})$ is an isomorphism on X. The extension V is preserved by T and T acts on the restriction of the local system V to D. As λ_1 is an isomorphism, T acts on V as a diagonal matrix $(\zeta^{-1}, 1, \dots, 1)$ with respect to a suitable base $e_0, e_1, \dots, e_n$. That $T e_i = e_i$ means that e_i comes from a holomorphic function $\bar{e}_i$ on Y. That $T e_0 = \zeta^{-1} e_0$ means that $e_0 = z_1 \cdot g_0$ with g_0 coming from a holomorphic function $\bar{g}$ on Y.

Comparing 6.17.3 and 6.7, the multivalued basis $(e_0, \dots, e_n)$ of V_Y shows that V_Y has strict exponents $(0, 1/r)$. Conversely if $(e_0, \dots, e_n)$ is a multi-valued basis of V_Y showing that V_Y has strict exponents $(0, 1/r)$, the functions e_i are holomorphic on X, with e_0 divisible by z_1, and the comparison of 6.17.1 and 6.7 proves (i).

Consider now the proof of (ii).

Using 6.12, it is easy to see that if V_Y has strict exponents $(\alpha/r, \beta/r)$, then V_X has strict exponents (α, β). Conversely, let V_X have strict exponents (α, β) and let $(e_0, \dots, e_n)$ be as in 6.7. The decomposition $V_X = V'_X \oplus V''_X$ of 6.7 (V'_X spanned by e_0, V''_X by the e_i, $i \neq 0$) is stable by T. The local systems of holomorphic functions $z_1^{-\beta} V'_X$ and $z_1^{-\alpha} V''_X$ extend to X (cf. 6.7). The action of T extends. The two local systems, restricted to D, embed into $\mathcal{O}_D$ (6.12). The actions of T on the restriction to D of $z^{-\beta} V'_X$

and $z^{-\alpha}V_X''$ are hence trivial, and the holomorphic functions $g_0 = z^{-\beta}e_0$, $g_i = z^{-\alpha}e_i(i \neq 0)$ are T-invariant, hence come from Y. Comparing 6.13 on X and Y gives 6.20.

6.21. Let X be an analytic variety of dimension n, D_1 and D_2 be two smooth divisors on X with transversal intersection, $D = D_1 + D_2$ and let V be a local system of holomorphic functions on $X - D$ of dimension $n + 1$.

Fix complex numbers $\alpha_i, \beta_i (i = 1, 2)$ with $\alpha_i - \beta_i \notin \mathbf{Z}$. Define

$$a_i = \exp(2\pi i \alpha_i) \ , \ b_i = \exp(2\pi i \beta_i).$$

PROPOSITION 6.22. *With the notations of 6.21, assume that*
(a) on $X - D$, $\lambda_1 : V \to jet_1(\mathcal{O})$ is an isomorphism
(b) V has strict exponents (α_1, β_1) along $D_1 - (D_1 \cap D_2)$ and (α_2, β_2) along $D_2 - (D_1 \cap D_2)$.

Then, in a neighborhood $U(p)$ of any point p of $D_1 \cap D_2$, V has a multivalued basis $(e_1, e_2, e_3, \ldots, e_{n+1})$ such that

$$\begin{aligned}
e_1 &= z_1^{\beta_1} z_2^{\alpha_2} g_1 \\
(6.22.1) \qquad e_2 &= z_1^{\alpha_1} z_2^{\beta_2} g_2 \\
e_i &= z_1^{\alpha_1} z_2^{\alpha_2} g_i \qquad (i \neq 1, 2)
\end{aligned}$$

with z_i a local equation for D_i, g_1 and g_2 holomorphic invertible in $U(p)$, and the $g_i (i \neq 1, 2)$ the projective coordinates of an etale map from $D_1 \cap D_2$ to $\mathbf{P}^{n-2}$.

PROOF. The question is local around $p \in D_1 \cap D_2$. We choose local coordinates $z_1, z_2, \ldots, z_n$ with z_i a local equation for $D_i (i = 1, 2)$, and we may assume that X is the polydisc $|z_i| < 1$. Let T_i be the monodromy around D_i. As T_1 and T_2 commute, they have a joint spectrum which is either

(a) $(b_1, a_2), (a_1, b_2), (a_1, a_2)$ with multiplicity $n - 1$.
(b) $(b_1, b_2), (a_1, a_2)$ with multiplicity n.

LEMMA 6.23. *A joint spectrum of type (b) for (T_1, T_2) is impossible.*

PROOF. If the joint spectrum is of type (b), V has a multivalued basis

$$\begin{aligned}
e_1 &= z_1^{\beta_1} z_2^{\beta_2} g_1 \\
e_i &= z_1^{\alpha_1} z_2^{\alpha_2} g_i \qquad (2 \leq i \leq n+1)
\end{aligned}$$

with univalued g's. By assumption, each g is holomorphic on $X - (D_1 \cap D_2)$, hence on X. Extend $V = \mathcal{O} \otimes V$ as a vector bundle V_X on X with the basis $z_1^{-\beta_1} z_2^{-\beta_2} \otimes e_1, z_1^{-\alpha_1} z_2^{-\alpha_2} \otimes e_i (2 \leq i \leq n+1)$. The map

$$\lambda_1 : V_X \to \mathrm{jet}_1(\mathcal{O})(\log\ D) := \mathcal{O} \oplus \Omega^1(\log\ D),$$

is invertible on $X - (D_1 \cap D_2)$, hence on the whole of X. Relative to the basis $(1, \frac{dz_1}{z_1}, \frac{dz_2}{z_2}, dz_i (3 \leq i \leq n))$ of $\mathcal{O} \oplus \Omega^1(\log\ D)$, and the above basis of V_X the columns of the matrix of λ_1, are

$$z_1^{-\beta_1} z_2^{-\beta_2} (1, z_1\partial_1, z_2\partial_2, \partial_3 \ldots \partial_n) z_1^{\beta_1} z_2^{\beta_2} g_1$$

$$z_1^{-\alpha_1} z_2^{-\alpha_2} (1, z_1\partial_1, z_2\partial_2, \partial_3 \ldots \partial_n) z_1^{\alpha_1} z_2^{\alpha_2} g_i \qquad (2 \leq i \leq n+1).$$

On $D_1 \cap D_2$, this reduces to

$$(1, \beta_1, \beta_2, \partial_3, \ldots, \partial_n) g_1$$

$$(1, \alpha_1, \alpha_2, \partial_3, \ldots, \partial_n) g_i \qquad (2 \leq i \leq n+1).$$

Solving $A + \beta_1 B + \beta_2 C = A + \alpha_1 B + \alpha_2 C = 0$, one sees that the first three rows are linearly dependent: a contradiction.

6.24 End of Proof 6.22. The joint spectrum of (T_1, T_2) is given by (a) and, reasoning as in 6.23, one sees that V has a multivalued basis (6.22.1) with holomorphic g's and that, on $D_1 \cap D_2$, the matrix of λ_1, with columns

$$(1, \beta_1, \alpha_2, \partial_3, \ldots, \partial_n) g_1$$

$$(1, \alpha_1, \beta_2, \partial_3, \ldots, \partial_n) g_2$$

$$(1, \alpha_1, \alpha_2, \partial_3, \ldots, \partial_n) g_i \qquad (3 \leq i \leq n+1)$$

is invertible. One has

$$\det \begin{pmatrix} 1 & \beta_1 & \alpha_2 \\ 1 & \alpha_1 & \beta_2 \\ 1 & \alpha_1 & \alpha_2 \end{pmatrix} = (\alpha_1 - \beta_1)(\alpha_2 - \beta_2) \neq 0;$$

the invertibility of λ_1 implies that of g_1, g_2 and of

$$D\begin{pmatrix} g_3, \cdots, & g_{n+1} \\ 1, & \partial_3, \cdots, & \partial_n \end{pmatrix}$$

By 6.3, this proves 6.22.

COROLLARY 6.25 (I). *If $\beta_1 = 0$, the local system $V'_{D_1 - D_1 \cap D_2}$ on $D_1 -$ $(D_1 \cap D_2)$ (6.7) is a rank one local system of invertible holomorphic functions (6.2). Its defining closed 1-form ν has a logarithmic pole and a residue α_2 along $D_1 \cap D_2$.*

(ii) If $\alpha_1 = 0$, the local system $V''_{D_1 - (D_1 \cap D_2)}$ on $D_1 - (D_1 \cap D_2)$ satisfies the equivalent conditions of 6.3 and has strict exponents (α_2, β_2) along $D_1 - (D_1 \cap D_2)$.

§7. CHARACTERIZATION OF
HYPERGEOMETRIC-LIKE LOCAL SYSTEMS

The results in this section are closely related to Terada's [Terada, 1973], which itself generalizes Picard [Picard, 1881]. A comparison with Terada's results will be given in (7.13).

We keep the notations $N, S, Q, Q^+, D_{s,t}$, and assume $N \geq 5$. For the case $N = 4$, see 7.14.

THEOREM 7.1. *Let V be a local system of holomorphic functions on Q. Assume that*

(i) *On Q, the map $\lambda_1 : V \to jet_1(\mathcal{O})$ of (6.3) is an isomorphism.*

(ii) *For suitable $\alpha_{s,t}, \beta_{s,t}$ ($s \neq t$ in S), with $\alpha_{s,t} - \beta_{s,t} \notin \mathbf{Z}$, V has strict exponents $(\alpha_{s,t}, \beta_{s,t})$ along $D_{s,t}$. Then, V is hypergeometric-like (4.7, 5.9).*

The μ relative to which V is hypergeometric-like is determined by

$$\beta_{s,t} - \alpha_{s,t} = 1 - \mu_s - \mu_t.$$

By 5.12, 6.6, 4.10, one has

COROLLARY 7.2. *V is uniquely determined by the $\alpha_{s,t}, \beta_{s,t}$.*

In 7.1(ii), we have assumed $\alpha_{s,t} - \beta_{s,t} \notin \mathbf{Z}$. The condition of strict exponents has not been defined otherwise.

The proof of 7.1 will occupy us until 7.12.

The exponents $\alpha_{s,t}, \beta_{s,t}$ depends only on the unordered pair $\{s, t\}$ of distinct elements of S. Sums over s, t will be sums over $\binom{S}{2}$.

7.3 Let $Q^{++}(S)$ or simply Q^{++} be the following partial compactification of Q: in $\mathbf{P}^{1^S}$, one takes the open subset U of S-uples $(x_s)_{s \in S}$ where no three of the x_s are equal, and at most two pairs of x_s are equal. For x in U, at least three of the x_s are distinct and the action of $PGL(2)$ on U is free. One defines $Q^{++} := U/PGL(2)$.

Fix $s, t \in S$, let $D^+_{s,t} \subset Q^{++}$ be the image of the divisor of U where $x_s = x_t$ and let S_{st} be the quotient of S obtained by identifying s and t. We keep writing s for the image of s in S_{st}. One has

$$(7.3.1) \qquad D^+_{s,t} = Q^+(S_{st}) - \bigcup_{u \neq s,t} D_{s,u}(S_{st}).$$

LEMMA 7.4. *If* $\beta_{s,t} = 0$, *then*

$$\sum_{\{s,t\} \cap \{u,v\} = \phi} \alpha_{u,v} = 0.$$

The sum is over all unordered pairs $\{u, v\} \subset S - \{s, t\}$.

PROOF. The local system V gives rise to a local system of invertible holomorphic functions $V'_{D^+_{s,t}}$ on $D^+_{s,t}$ (6.6 and 6.12). The defining 1-form ν on $D_{s,t} = Q(S_{st})$ has a simple pole with residue $\alpha_{u,v}$ along $D_{u,v}(S_{st}) \subset Q^+(S_{st})$ (6.25). Applying 2.11 to $D^+_{u,v}(S_{st})$, we get the required identity from 2.6 and 3.4.

PROPOSITION 7.5. *For* V *as in 7.1, one has*

(i) $\beta_{s,t} = \sum_{\{u,v\}} \alpha_{u,v}$ *(sum over* $\binom{S - \{s,t\}}{2}$*) and*

(ii) $\sum_{\{s,t\}} \alpha_{s,t} = 1$ *(sum over* $\binom{S}{2}$*).*

If one defines $\mu_s := \sum_t \alpha_{s,t}$, the equation (ii) amounts to

$$\sum \mu_s = 2.$$

Let $A_{u,v}$ be any family of numbers dependng on a two element subset of $S : A_{u,v} = A_{v,u}$ and $A_{u,u}$ undefined (or $= 0$ if one prefers). One has

$$\sum_{\{s,t\} \cap \{u,v\} = \phi} A_{u,v} = \frac{1}{2} \Big(\sum_{u,v} A_{u,v} - \sum_{u=s} A_{u,v}$$

$$- \sum_{u=t} A_{u,v} - \sum_{v=s} A_{u,v}$$

$$- \sum_{v=t} A_{u,v} + A_{s,t} + A_{t,s} \Big)$$

$$(7.5.1) \qquad = \sum_{\{u,v\}} A_{u,v} - \sum A_{s,u} - \sum A_{t,u} + A_{s,t}.$$

Granting (ii), (i) can hence be rewritten

$$\beta_{s,t} = 1 - \mu_s - \mu_t + \alpha_{s,t}, \text{ i.e.,}$$
$$\beta_{s,t} - \alpha_{s,t} = 1 - \mu_s - \mu_t.$$

The proposition amounts to say that there is a hypergeometric-like local system with the same exponents as V.

PROOF. Fix s, t. Let f be a multivalued function of the form $\prod f_i^{\alpha_i}$ with $f_i \in \mathcal{O}^*(Q)$ and $\alpha_i \in \mathbf{C}$. We choose f so that the valuation $v_{s,t}(f)$ along $D_{s,t}$ is $-\beta_{st}$. This is possible: the only constraint on the valuations of f is that for each u, $\sum_v v_{u,v}(f) = 0$.

Lemma 7.4 applies to the twisted local system fV:

$$\sum_{\{s,t\} \cap \{u,v\} = \phi} \alpha_{u,v} + v_{u,v}(f) = 0.$$

The sum of the $v_{u,v}$ can be rewritten by (7.5.1) as

$$0 - 0 - 0 + v_{s,t}(f) = -\beta_{s,t}$$

and (i) follows.

On Q^+, let us consider the vector bundle V of (6.7) and its maximal exterior power $\det V$. The connection ∇ of V on Q induces one on $\det V$. Let z be a local equation for $D_{s,t}$ near $p \in D_{s,t}$. Near p, $\det V$ has a basis

$$(z^{-\beta_{s,t}} \otimes e_0) \wedge (z^{-\alpha_{s,t}} \otimes e_1) \wedge \cdots \wedge (z^{-\alpha_{s,t}} \otimes e_n)$$
$$= z^{-\beta_{s,t} - (N-3)\alpha_{s,t}} \cdot \text{ multivalued horizontal section.}$$

The residue of the connection is hence $-(\beta_{st} + (N-3)\alpha_{st})$ and by 2.10, 2.6, one has in $\text{Pic}(Q^+) \otimes \mathbf{C}$

$$\det V = \sum (\beta_{s,t} + (N-3)\alpha_{s,t}) cl(D_{s,t})$$

i.e., for each $s \in S$ (cf. (3.1.4))

$$d_s(\det V) = \sum_t (\beta_{s,t} + (N-3)\alpha_{s,t}).$$

By the assumption on strict exponents, $\mathcal{V}$ is isomorphic to

$$\mathcal{O} \oplus \Omega^1 (\log \sum D_{s,t}),$$

whose d_s is $N - 3$. This gives for each s

$$\sum_t (\beta_{s,t} + (N - 3)\alpha_{s,t}) = N - 3.$$

By (i), the sum over $\beta_{s,t}$ can be rewritten

$$(N - 3) \sum_{\{u,v\} \subset S - \{s\}} \alpha_{u,v}$$

as each $\alpha_{u,v}$ occurs in the sum for $\beta_{s,t}$ for $t \ne s, u, v$. Assertion (ii) follows.

7.6 Fix V as in 7.1 and choose $s, t \in S$. Let, as in 5.11, $\Xi_{s,t}$ be the differential operator on Q which transform a function f into the function whose pull-back $\Xi_{s,t}(f) \circ \pi$ on H is given by

$$\Xi_{s,t}(f) \circ \pi = \prod_{a,b} F_{a,b}^{\alpha_{a,b}} \cdot F_{s,t} \Delta_{s,t} (\prod_{a,b} F_{a,b}^{-\alpha_{a,b}} \cdot f \circ \pi).$$

The restriction of $\Xi_{s,t}$ to V extends by $\mathcal{O}$-linearity to a linear form on $\mathcal{V}$.

We extend the vector bundle $\mathcal{V}$ to a vector bundle on Q^+ as in 6.7. We write $\bar{\mathcal{V}}$ for the extended bundle.

Choose a divisor $D_{u,v}$. Near $D_{u,v}$, we have a decomposition $V = V' \oplus V''$ as in 6.6, and it induces a decomposition $\bar{\mathcal{V}} = \bar{\mathcal{V}}' \oplus \bar{\mathcal{V}}''$.

PROPOSITION 7.7. *The linear form $\Xi_{s,t}$ extends as follows on a neighborhood of $D_{u,v} \subset Q^+$:*

(i) $\{s,t\} \cap \{u,v\} = \phi$: *holomorphic extension* $\bar{\mathcal{V}} \to \mathcal{O}$.

(ii) $\#(\{s,t\} \cap \{u,v\}) = 1$: *holomorphic extension*

$$\bar{\mathcal{V}}'(-D_{u,v}) \oplus \bar{\mathcal{V}}'' \to \mathcal{O}.$$

(iii) $\{s,t\} = \{u,v\}$: *holomorphic extension* $\bar{\mathcal{V}}(D_{u,v}) \to \mathcal{O}$.

PROOF. It suffices to prove (i) (ii) (iii) after pulling back from Q^+ to H^+. We denote the pullback of $V, V', V'', \bar{V}, \bar{V}', \bar{V}''$ and the linear form $\Xi_{s,t}$ by the same letter. The pullback of the divisor $D_{u,v}$ is $F_{u,v} = 0$.

The vector bundle $\bar{\mathcal{V}}''$ is spanned near $D_{u,v}$ by sections $F_{u,v}^{-\alpha_{u,v}} \otimes e$ with e a multivalued section of V''. The image of this section of $\bar{\mathcal{V}}''$ by $\Xi_{s,t}$ is

$$F_{u,v}^{-\alpha_{u,v}} \cdot \prod F_{a,b}^{\alpha_{a,b}} \cdot F_{s,t}\Delta_{s,t} \left(\prod F_{a,b}^{-\alpha_{a,b}} \cdot e \right) =$$

$$\prod_{\{a,b\} \neq \{u,v\}} F_{a,b}^{\alpha_{a,b}} \cdot F_{s,t}\Delta_{s,t} \left(\prod_{\{a,b\} \neq \{u,v\}} F_{a,b}^{-\alpha_{a,b}} \cdot F_{u,v}^{-\alpha_{u,v}} e \right).$$

The product of the $F_{a,b}^{\alpha_{a,b}}$ for $\{a,b\} \neq \{u,v\}$ is holomorphic invertible near $D_{u,v}$, $F_{u,v}^{-\alpha_{u,v}} e$ is holomorphic, and one concludes

$$\Xi_{s,t}(F_{u,v}^{-\alpha_{u,v}} \otimes e) = F_{s,t} \cdot \text{ holomorphic near } D_{u,v}.$$

This agrees with 7.7.

The line bundle $\bar{\mathcal{V}}'$ is spanned over $D_{a,b}$ by a section $F_{u,v}^{-\beta_{u,v}} \otimes e$ with e a multivalued section of V'. The image of this section of $\bar{\mathcal{V}}'$ by $\Xi_{s,t}$ is

$$\prod_{\{a,b\} \neq \{u,v\}} F_{a,b}^{\alpha_{a,b}} \cdot F_{u,v}^{\alpha_{u,v} - \beta_{u,v}} \cdot F_{s,t}\Delta_{s,t}(F_{u,v}^{\beta_{u,v} - \alpha_{u,v}}.$$

$$\prod_{\{a,b\} \neq \{u,v\}} F_{a,b}^{-\alpha_{a,b}} \cdot F_{u,v}^{-\beta_{u,v}} e)$$

We have $\prod_{\{a,b\} \neq \{u,v\}} F_{a,b}^{\alpha_{a,b}}$ holomorphic invertible near $D_{u,v}$, $\beta_{u,v} - \alpha_{u,v} = 1 - \mu_u - \mu_v$ and $F_{u,v}^{-\beta_{u,v}} e$ is holomorphic near $D_{u,v}$. We can rewrite

$$\Xi_{s,t}(F_{u,v}^{-\beta_{u,v}} \otimes e) = \text{ holomorphic } \cdot F_{u,v}^{-(1-\mu_u-\mu_v)} \cdot F_{s,t}\Delta_{s,t}(F_{u,v}^{1-\mu_u-\mu_v} f)$$

with f holomorphic and homogeneous in each pair $(h_{s,1}, h_{s,2})$. The degree of homogeneity of f is $-\mu_s$ for $s \neq u, v$, $\mu_u - 1$ for $s = v$, and $\mu_v - 1$ for $s = u$, inasmuch as $F_{u,v}^{1-\mu_u-\mu_v} f = \prod F_{a,b}^{-\alpha_{a,b}} e$ is homogeneous of degree $-\mu_s$.

If $\{s,t\} \cap \{u,v\} = \phi$, $\Delta_{s,t}(F_{u,v}^A f) = F_{u,v}^A \Delta_{s,t}(f)$ and (i) follows.

If $u = s, v \neq t$, one has

$$\Delta_{s,t}(F_{u,v}^A f) = (\partial_{s,1}\partial_{t,2} - \partial_{s,2}\partial_{t,1})((h_{s,1}h_{v,2} - h_{s,2}h_{v,1})^A f)$$
$$= F_{u,v}^A \Delta_{s,t} f + A F_{u,v}^{A-1}(h_{v,2}\partial_{t,2} + h_{v,1}\partial_{t,1})(f)$$

and (ii) follows.

To handle the case $u = s, v = t$, we will use

LEMMA 7.8. *If f is homogeneous of degree A_1 in $(h_{s,1}, h_{s,2})$, of degree A_2 in $(h_{t,1}, h_{t,2})$, and if $A = -1 - A_1 - A_2$, then*

$$\Delta_{s,t}(F_{s,t}^A f) = F_{s,t}^A \Delta_{s,t} f.$$

PROOF. One has

$$\partial_{t,2}[F_{s,t}^A f] = F_{s,t}^A \partial_{t,2} f + A F_{s,t}^{A-1} \cdot \partial_{t,2} F_{s,t} \cdot f$$

$$\begin{aligned}
\partial_{s,1} \partial_{t,2}[F_{s,t}]^A f &= F_{s,t}^A \partial_{s,1} \partial_{t,2} f + A F_{s,t}^{A-1} \partial_{s,1} F_{s,t} \partial_{t,2} f \\
&+ A F_{s,t}^{A-1} \partial_{t,2} F_{s,t} \partial_{s,1} f + A F_{s,t}^{A-1} \partial_{s,1} \partial_{t,2} F_{s,t} \cdot f \\
&+ A(A-1) F_{s,t}^{A-2} \partial_{s,1} F_{s,t} \cdot \partial_{t,2} F_{s,t} \cdot f \\
&= F_{s,t}^A \partial_{s,1} \partial_{t,2} f \\
&+ A F_{s,t}^{A-1}(h_{t,2} \partial_{t,2} + h_{s,1} \partial_{s,1} + 1) f \\
&+ A(A-1) F_{s,t}^{A-2} h_{s,1} h_{t,2} f.
\end{aligned}$$

Similarly

$$\begin{aligned}
\partial_{s,2} \partial_{t,1}[F_{s,t}^A f] &= F_{s,t}^A \partial_{s,2} \partial_{t,1} f + A F_{s,t}^{A-1}(-h_{s,2} \partial_{s,2} - h_{t,1} \partial_{t,1} - 1) \\
&+ A(A-1) F_{s,t}^{A-2} h_{t,1} h_{s,2} f \\
\Delta_{s,t}[F_{s,t}^A f] &= F_{s,t}^A \Delta_{s,t} f \\
&+ A F_{s,t}^{A-1}(h_{s,1} \partial_{s,1} + h_{s,2} \partial_{s,2} + h_{t,1} \partial_{t,1} \\
&+ h_{t,2} \partial_{t,2} + 2) f \\
&+ A(A-1) F_{s,t}^{A-1} f \\
&= F_{s,t}^A \Delta_{s,t} f + A F_{s,t}^{A-1}(A_1 + A_2 + 2 + (A-1)) f
\end{aligned}$$

and the second term vanishes if $A_1 + A_2 + A + 1$ does.

We can now conclude the proof of 7.7. For $u = s$, $v = t$, 7.8 applies with $A = 1 - \mu_u - \mu_v$, $A_1 = \mu_u - 1$, $A_2 = \mu_v - 1$, hence

$$\Xi_{s,t}(F_{u,v}^{-\beta_{u,v}} \otimes e) = F_{s,t} \cdot \text{holomorphic},$$

as required.

7.9 The morphism λ_1 extends by assumption to an isomorphism on Q^+ from $\bar{V}$ to $\text{jet}_1(\mathcal{O})(\log \Sigma D_{ab}) = \mathcal{O} + \Omega^1(\log \Sigma D_{ab})$. On the divisor $D = D_{u,v}$, we have two natural maps from $\text{jet}_1(\mathcal{O})(\log D)$ to $\mathcal{O}_D$: the projection from $\mathcal{O}$ to $\mathcal{O}_D$ and the residue map on $\Omega^1(\log D)$.

LEMMA 7.10. *Near $D_{u,v}$, the intersection $\mathcal{O}(-D_{u,v}) \oplus \Omega^1$ of the kernels of the two maps of 7.9: $\mathrm{jet}_1(\log D_{u,v}) \to \mathcal{O}_{D_{u,v}}$ is the image by λ_1 of*

$$\ker(\lambda : \bar{\mathcal{V}}' \to \mathcal{O}_{D_{u,v}}) \oplus \ker(\lambda : \bar{\mathcal{V}}'' \to \mathcal{O}_{D_{u,v}}).$$

PROOF. Let λ' be the composite $\bar{\mathcal{V}} \to \bar{\mathcal{V}}' \xrightarrow{\lambda} \mathcal{O} \to \mathcal{O}_D$ and λ'' the composite $\bar{\mathcal{V}} \to \bar{\mathcal{V}}'' \xrightarrow{\lambda} \mathcal{O} \to \mathcal{O}_D$. The two maps 7.9 are respectively $\lambda' + \lambda''$ and $\beta_{u,v}\lambda' + \alpha_{u,v}\lambda''$.

The kernel $\ker(\lambda : \bar{\mathcal{V}}' \to \mathcal{O}_{D_{u,v}})$ is simply $\bar{\mathcal{V}}'(-D_{u,v})$. We deduce from 7.7 and 7.10:

COROLLARY 7.11. *The linear form $\Xi_{s,t} \circ \lambda_1^{-1}$ on $\mathrm{jet}_1(\mathcal{O})$, restricted to Ω^1, extends as follows on a neighborhood of $D_{u,v} \subset Q^+$:*

  (i) *$\{u,v\} \cap \{s,t\} = \phi$:*
     holomorphic extension $\Omega^1(\log D_{u,v}) \to \mathcal{O}$.
  (ii) *$\#(\{u,v\} \cap \{s,t\}) = 1$:*
     holomorphic extension $\Omega^1 \to \mathcal{O}$.
  (iii) *$\{u,v\} = \{s,t\}$:*
     holomorphic extension $\Omega^1(\log D_{u,v})(D_{u,v}) \to \mathcal{O}$.

7.12 Proof of 7.1 (concluded)

A linear form on Ω^1 is a vector field. By 7.11, the vector field corresponding to the linear form $\Xi_{s,t} \circ \lambda^{-1}$ restricted to Ω^1 is a vector field on Q which extends to a vector field τ on Q^+. In addition, τ is tangent to the $D_{u,v}$ for $\{u,v\} \cap \{s,t\} = \phi$ and, near $D_{s,t}$, is the product of a local equation for $D_{s,t}$ by a vector field tangent to $D_{s,t}$.

LEMMA. *A vector field as above vanishes identically.*

PROOF. Fix a, b, c in S distinct from s and t and define $S_1 = S - \{a,b,c\}$. We identify Q with the S-uples of points in $\mathbf{P}^1$ with $x_a = 0, x_b = 1, x_c = \infty$ and compactify it by $(\mathbf{P}^1)^{S_1}$. If we remove from $\mathbf{P}^{1^{S_1}}$ a subset of codimension 2, we get $Q^+ - D_{a,b} - D_{b,c} - D_{a,c}$. The vector field τ hence extends to $\mathbf{P}^{1^{S_1}}$. The tangent bundle to $\mathbf{P}^{1^{S_1}}$ is the sum of the pullback by the pr_i of the tangent bundle H of $\mathbf{P}^1$, and τ is the sum of vector fields τ_i $(i \in S_1)$, $\tau_1 \in \Gamma(\mathbf{P}^{1^{S_1}}, pr_i^* H)$. For any coherent sheaf $\mathcal{F}$ on $\mathbf{P}^1$, one has $\Gamma(\mathbf{P}^1, \mathcal{F}) \xrightarrow{\sim} \Gamma(\mathbf{P}^{1^{S_1}}, pr_1^* \mathcal{F})$. This results, for instance, from the Künneth formula and the fact that $\Gamma(\mathbf{P}^1, \mathcal{O}) = \mathbf{C}$. In particular, τ_i comes from a vector field $\bar{\tau}_i$ on $\mathbf{P}^1$. Inasmuch as the vector field τ vanishes on $D_{s,t}$, so also does each τ_i. Since pr_i maps $D_{s,t}$ onto P^1, $\bar{\tau}_i$ vanishes. The lemma follows.

We conclude that $\Xi_{s,t} \circ \lambda_s^{-1}$ factors through the quotient $\mathcal{O}$ of $\mathrm{jet}_1(\mathcal{O})$, i.e., that on $\mathcal{V}$,

$$\Xi_{s,t} = f\lambda$$

for a suitable function f. On Q^+, near any $D_{u,v}$, λ maps $\bar{\mathcal{V}}'$ onto $\mathcal{O}$ and one concludes from 7.7 that f is holomorphic on Q^+ (hence a constant by 3.2) and vanishes on $D_{s,t}$. Thus $f = 0$. One concludes that $\Xi_{s,t} = 0$ and that the local sections v of V satisfy the differential equations $\Xi_{s,t}(v) = 0$. Those equations are enough to express each second derivative in terms of first derivatives by (5.9) and thus V is uniquely determined by its exponents. It coincides with the hypergeometric-like system with the same exponents (5.11) and this concludes the proof of 7.1.

7.13 In [Picard, 1881] Picard considered functions of two variables of the type considered here and proved a uniqueness theorem for them. Terada [Terada, 1973] proved a theorem similar to 7.1, under the additional assumption that the exponents α, β are those of Lauricella local system. If we add to the hypothesis of 7.1 that the exponents (α, β) are the exponents of a hypergeometric-like local system, then we can derive the conclusion of 7.1 from Terada's theorem by making an additional twist.

Other differences between our theorems:

(a) Instead of a "strict exponent" condition, Terada imposes a condition (6.6.1). The "strict exponent" condition is recovered from his assumptions as follows. With the notations of 6.7, one considers

$$\lambda_1 : \mathcal{V} \to \mathrm{jet}_1(\mathcal{O})(\log \Sigma D_{s,t})$$

and $\det(\lambda_1)$ from $\overset{N-2}{\bigwedge} \mathcal{V}$ to $\overset{N-2}{\bigwedge} \mathrm{jet}_1(\mathcal{O})(\log \Sigma D_{s,t})$. Terada's proof requires the unstated assumption (which he omits) that $\det(\lambda_1)$ is not identically zero. The assumption on the exponents ensures that the line bundles

$$\overset{N-2}{\bigwedge} \mathcal{V} \text{ and } \overset{N-2}{\bigwedge} \mathrm{jet}_1(\mathcal{O})(\log \Sigma D_{s,t})$$

are isomorphic (cf. the proof of 7.5). The divisor of $\det(\lambda_1)$ is hence the divisor of a function. Since holomorphic functions on Q^+ are constant, $\det(\lambda_1)$ is invertible, λ_1 is too, and the strict exponent condition 6.9 is met.

(b) He fixes $a, b, c \in S$ and imposes conditions only along some of the divisors $D_{s,t}$; at $D_{a,b}, B_{a,c}, D_{b,c}$ no condition is imposed. This is a strengthening of 7.1. It leads us to the following question:

Question Fix any system of numbers ν_s $(s \in S)$ with $0 < \nu_s < 1$, $\Sigma \nu_s = 2$. In 7.1, can one weaken the assumption (ii) to read: "V has strict exponents along $D_{s,t}$ whenever $\nu_s + \nu_t < 1$"?

7.14. The case $N = 4$. The case $N = 4$ is that of classical hypergeometric functions. By Riemann, on $\mathbf{P}^1 - \{0, 1, \infty\}$ there is a unique rank 2 local system of holomorphic functions, with given strict exponents $(\alpha_0, \beta_0)(\alpha_1, \beta_1)(\alpha_\infty, \beta_\infty)$ at $0, 1, \infty$, assuming the equivalent conditions:

(7.14.1) The sum of the exponents is 1,

(7.14.2) $\lambda_1 : V \to \mathrm{jet}_1(\mathcal{O})$ is an isomorphism.

Only the unordered pairs $(\alpha_0, \beta_0)(\alpha_1, \beta_1)(\alpha_\infty, \beta_\infty)$ matter.

The space Q^+ is $\mathbf{P}^1$ with $0, 1$ and ∞ doubled: it is non Hausdorff; for each partition $\{a, b\}\{c, d\}$ of S, the points $D_{a,b}$ and $D_{c,d}$ lie above the same point of $\mathbf{P}^1$. On the $D_{s,t} \in Q^+$, one should consider systems of exponents $(\alpha_{s,t}, \beta_{s,t})$ such that for each partition $\{a, b\}, \{c, d\}$ of S one has $\beta_{c,d} = \alpha_{a,b}$. The condition 7.14.1 then reads $\Sigma \alpha_{s,t} = 1$ (as in 7.5(ii)) while 7.5(i) holds by definition.

When the exponents at $0, 1, \infty$ are given, there is more than one way to assign the $\alpha_{s,t}$. To go from one to another, one can renumber the $\alpha_{s,t}$ by a Vierergruppe element (not surprising, since the Vierergruppe acts trivially on Q^+) as well as replace each $\alpha_{s,t}$ by the corresponding $\beta_{s,t}$. On the μ's, this corresponds respectively to applying a Vierergruppe element and replacing $(\mu_s)_{s \in S}$ by $(1 - \mu_s)_{s \in S}$.

7.15 Let H_2 be a subgroup of the symmetric group $\Sigma(S)$ and $_2Q'$ be the open subset of Q on which H_2 acts freely. If $_2\alpha$ is a family of complex numbers, indexed by $\binom{S}{2}$, H_2-invariant, and of sum 1, the hypergeometric-like local system $L(_2\alpha)$ is H_2-invariant, hence descends to $L(_2\alpha)/H_2$ on $_2Q'/H_2$. We will use 7.1 to prove that pullback of suitable local systems of holomorphic functions $L(_2\alpha)/H_2$ by suitable rational maps $\psi : Q \to Q/H_2$ are hypergeometric-like. The setting will be as follows.

(a) H_2 and $_2Q'$ are as above; and $_2Q''$ is a Zariski open, dense and H_2-stable subset of Q^+. Let $_2I$ be the set of irreducible components of $_2Q''/H_2 -_2 Q'/H_2$ which are of codimension one in $_2Q''/H_2$. In our setting, codimension 2 phenomena will be irrelevant: we may shrink $_2Q''$ by removing a codimension 2 subset and assume that $_2Q''/H_2$ is smooth and that $_2Q''/H_2 -_2 Q'/H_2$ is a disjoint union of smooth irreducible divisors.

Fix $D \in {}_2I$ and an irreducible component $\tilde{D}$ of its inverse image in $_2Q''$. We say that D is at infinity (resp. at finite distance or finite) if $\tilde{D}$ is dense in $D_{s,t}$ (resp. in a codimension one component of the fixed point set of some $h \in H_2$ on Q). We denote by $\tilde{r}_2(D)$ the ramification index along $\tilde{D}$ of the quotient map $_2\tau :_2 Q'' \to_2 Q''/H_2$. By abuse of notation, if $_2Q'' \cap D_{u,v}$ is dense in $D_{u,v}$, we will keep writing $D_{u,v}$ for the divisor $_2Q'' \cap D_{u,v}$ of $_2Q''$.

(b) $_1Q''$ is a Zariski open subset of Q^+ whose complement is of codimension ≥ 2 and $\psi : {}_1Q'' \to {}_2Q''/H_2$ is a quasi-finite map (i.e., with finite fibers). (Note the asymmetry: we have not required the complement of $_2Q''$ in Q^+ to be of codimension ≥ 2). Let $_1I$ be the set of irreducible divisors D of $_1Q''$ which are either the trace of a $D_{s,t}$ (divisors at infinity) or a ramification divisor for ψ, which is not the trace of a $D_{s,t}$ (divisor at finite distance). We denote by r_ψ the ramification index of the map ψ along D. By abuse of notation, we will write $D_{s,t}$ for $_1Q'' \cap D_{s,t}$.

(c) We assume that the union of the divisors in $_1I$ is the full inverse image of the union of the divisors in $_2I$. If D and $\psi(D)$ are both finite, we assume that

$$r_\psi(D) = \tilde{r}_2(\psi(D)) \qquad (D \text{ and } \psi(D) \text{ finite}).$$

This means that, in a neighborhood of a general point of D, ψ can be lifted to an etale map to $_2Q''$.

We also assume that

(i) if D is at infinity and $\psi(D)$ is at finite distance, then $r_\psi(D)/\tilde{r}_2(\psi(D))$ is not an integer.

(ii) if D is at finite distance and $\psi(D)$ is at infinity, then $\tilde{r}_2(\psi(D))/r_\psi(D)$ is not an integer.

Let $_2\alpha$ be a family of complex numbers, indexed by $\begin{pmatrix} S \\ 2 \end{pmatrix}$, with sum 1. As usual, one defines

$$_2\mu_u := \sum_v {_2\alpha_{u,v}} \text{ and } _2\beta_{s,t} := \alpha_{s,t} + (1 - _2\mu_s - _2\mu_t).$$

We will later explain a variant, where only $_2\mu$ is given. We assume that

(d) $_2\alpha$ is H_2-invariant.

(e) If $D \in_1 I$ is finite, with $\psi(D)$ at infinity and a quotient of $D_{u,v}$, then

$$(_2\alpha_{u,v}, _2\beta_{u,v}) = \frac{\tilde{r}_2(\psi(D))}{r_\psi(D)}(0, 1).$$

Assumption 7.15(c) (ii) assures that $_2\alpha_{u,v} - _2\beta_{u,v} \notin \mathbf{Z}$.

(f) non integrality: If $\psi(D_{s,t})$ is a quotient of $D_{u,v}$; then $_2\mu_u + _2\mu_v \notin \mathbf{Z}$ and $\frac{r_\psi(D_{s,t})}{\tilde{r}_2(\psi(D_{s,t}))}(1 - _2\mu_u - _2\mu_v) \notin \mathbf{Z}$.

We will show, as a consequence of 7.1:

COROLLARY 7.16. *Under the assumptions of 7.15, the local system of holomorphic functions* $\psi^*(L(_2\alpha)/H_2)$ *on* $\psi^{-1}(_2Q''\cap_2Q')$ *extends to a hypergeometric-like local system of holomorphic functions on Q. Its exponents* $(_1\alpha_{s,t}, _1\beta_{s,t})$ *are given by*

(a) If $\psi(D_{s,t})$ is at finite distance: $(_1\alpha_{s,t}, _1\beta_{s,t}) = \frac{r_\psi(D_{s,t})}{\tilde{r}_2(\psi(D_{s,t}))} \cdot (0, 1)$

(b) If $\psi(D_{s,t})$ is at infinity, quotient of $D_{u,v}$:

$$(_1\alpha_{s,t}, _1\beta_{s,t}) = \frac{r_\psi(D_{s,t})}{\tilde{r}_2(\psi(D_{s,t}))}(_2\alpha_{u,v}, _2\beta_{u,v}).$$

PROOF. We first prove that on $Q \cap_1 Q''$, $\psi^* L(_2\alpha)/H_2$ extends to a local system with the property

(*) a local basis forms projective coordinates for an etale map to projective space.

For x outside of the divisors in $_1I$, ψ is etale at x, $_2Q'' \to_2 Q''/H_2$ etale above $\psi(x)$, property (*) at x results from property (*) for $L(_2\alpha)$. Consider now a divisor D of $_1I$ at finite distance. If $\psi(D)$ is finite, near any $x \in D$, the map ψ lifts to an etale map to $_2Q$ by 7.15(c), and again (*) at x follows from property (*) for $L(_2\alpha)$. If $\psi(D)$ is infinite, a quotient of $D_{u,v}$, the assumption 7.15c (ii) assures us that $L(_2\alpha)$ has strict exponents at $D_{u,v}$;

assumption 7.15(e) assures us, by 6.20 that $\psi^* L(_2\alpha)/H_2$ extends across D and that (*) holds for the extension.

The complement of $Q \cap Q''$ in Q is of codimension ≥ 2. Consequently, the local system $\psi^*(L_2\alpha)/H_2$ extends to Q and property (*), which by 6.3 is expressible as a non-vanishing condition, holds for the extension.

In view of Theorem 7.1, it remains only to prove that the extended local system on Q has strict exponents along each $D_{s,t}$. Let D be a $D_{s,t}$. If $\psi(D_{s,t})$ is at finite distance, it follows from 6.20 (ii) and the non-integrality (c) of $r_\psi(D)/\tilde{r}_2(\psi(D))$ that $\psi^* L(_2\alpha)/H_2$ has strict exponents along $D_{s,t}$ given by 7.16 (a). If $\psi(D_{s,t})$ is at infinity, it follows from 6.20 (ii) and the second non-integrality assumption 7.15 (f) that it has strict exponent along $D_{s,t}$ given by 7.16 (b). The Corollary 7.16 now follows from 7.1.

Remark 7.17 (i) In the space $\mathbf{C}\binom{S}{2}$, the conditions on $_2\alpha$: "sum 1", H_2-invariance, and 7.15 (e) define an affine subspace A. The non-integrality condition 7.15 (f) on $_2\mu$ imposes a condition that either never holds on A or holds in a dense subset of A. For the maps ψ that we will consider, they will hold in a dense subset. The conclusion of 7.16, in the form: "$\psi^*(L(_2\alpha)/H_2)$ is the restriction to $\psi^{-1}(_2Q'' \cap_2 Q')$ of $L(_1\alpha)$, with $_1\alpha$ given by 7.16 (a)(b)" then holds without assuming the non-integrality 7.15 (f). More explicitly, for $_2\alpha$ in a dense subset of A, $\psi^*(L(_2\alpha)/H_2)$ is the restriction of $L(_1\alpha)$, to a fixed open dense subset of Q with $_1\alpha$ an affine linear function of $_2\alpha$; hence the same assertion holds for all $_2\alpha$ in A by continuity.

7.17 (ii) The local system of holomorphic functions $L(_2\alpha)/H_2$ is uniquely determined from its pullback by ψ. It follows that the affine map $_2\alpha \mapsto_1 \alpha$ defined by 7.16 (a)(b) on A is injective.

7.18 There is a projective variant of 7.16 which involves hypotheses on ψ and $_2\mu$ only.

Let $_2\mu$ be a family of complex numbers, indexed by S, with sum 2. We assume:

(d') $_2\mu$ is H_2-invariant.

By 4.12 or 5.8, there exists a family of numbers $_2\alpha_{s,t}$ indexed by $\binom{S}{2}$ for which

$$2\mu_s = \sum_t {}_2\alpha_{s,t}.$$

These equations being linear, one gets by averaging an H_2-invariant family. We choose an H_2-invariant $_2\alpha$ related to $_2\mu$ in this way.

(e') if $D \in_1 I$ is finite, with $\psi(D)$ at infinity, quotient of $D_{u,v}$, then

$$1 -_2 \mu_u -_2 \mu_v = \frac{\tilde{r}_2(\psi(D))}{r_\psi(D)}$$

(f') the same non-integrality conditions on $_2\mu_u$ as in 7.15 (f).

COROLLARY 7.19. *Under the assumptions 7.15 (a)(b)(c) and 7.18 on ψ and $_2\mu$, there exists on $\psi^{-1}(_2Q'' \cap_2 Q') \subset Q$-(finite divisors) a multivalued function f, of the form $f = \prod_i f_i^{a_i}$ with f_i holomorphic (in fact, algebraic) invertible and $a_i \in \mathbf{C}$, such that $f \cdot \psi^* L(_2\alpha)/H_2$ extends to a hypergeometric-like local system of holomorphic functions on Q. Its associated $_1\mu$ is given by:*

(a) *If $\psi(D_{s,t})$ is finite:* $1 -_1\mu_s -_1\mu_t = \frac{r_\psi(D_{s,t})}{\tilde{r}_2(\psi(D_{s,t}))}$.

(b) *If $\psi(D_{s,t})$ is at infinity, quotient of $D_{u,v}$:*

$$1 -_1\mu_s -_1\mu_t = \frac{r_\psi(D_{s,t})}{\tilde{r}_2(\psi(D_{s,t}))}(1 -_2\mu_u -_2\mu_v).$$

PROOF. We repeat the proof of 7.16. This time, if D is a finite divisor with $\psi(D)$ at infinity and a quotient of $D_{u,v}$, if z is a local equation of D, and if $\gamma(D) = \frac{r_\psi(D)}{\tilde{r}_2(\psi(D))} \cdot {}_2\alpha_{u,v}$, then the local system $z^{-\gamma(D)}\psi^*(L(_2\alpha)/H_2)$ extends across D, still satisfying 7.16 (*) by (6.11) and (6.12). We know that $Pic(Q) = 0$. For each finite divisor D, there hence exists a global equation z_D of $\bar{D}$ in Q. One takes $f := \prod z_D^{-\gamma(D)}$ and, as in 7.16, 7.19 follows from 7.1.

Remarks 7.20 (i) The analogue of the continuity argument 7.17 holds for 7.19, i.e., the non-integrality assumption 7.18 (f') can be dropped for $_2\mu$ in an affine family whose generic element satisfies 7.18 (f').

(ii) For the maps ψ which we will consider, each $_2\mu$ as in 7.18 comes from an $_2\alpha$ satisfying 7.15 (e), so that 7.19 is an immediate consequence of 7.16.

(iii) It follows from 7.19 that the equations 7.19 (a) (b) have a solution $_1\mu$, and that the sum of the $_1\mu_s$ is 2.

7.21 Suppose now that H_1 is a subgroup of the symmetric group $\Sigma(S)$, that $_1Q''$ is H_1-stable and that ψ is H_1-invariant; ψ factors through the quotient map $_1\tau : {}_1Q'' \to {}_1Q''/H_1$:

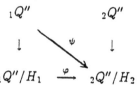

Let $_1Q'$ be the open subset of Q on which H_1 acts freely. With the assumptions of 7.15 and $_1\alpha$ as in 7.16, $L(_1\alpha)$ descends to $L(_1\alpha)/H_1$ on $_1Q'/H_1$ and, on a suitable dense open set, one has

$$L(_1\alpha)/H_1 = \varphi^* L(_2\alpha)/H_2.$$

Let U be a (dense) Zariski open subset of $_2Q''/H_2$, over which φ is an etale covering, and with $\varphi^{-1}(U) \subset {}_1Q'/H_1$. Let us fix a base point $0 \in \varphi^{-1}(U)$, and its image $\varphi(0)$ in U. The monodromy actions of π_1 on the local systems $L(_i\alpha)/H_i$ ($i = 1, 2$) are related by the commutative diagram

$$\begin{array}{ccc} \pi_1(\varphi^{-1}(U), 0) & \longrightarrow & \pi_1(U, \varphi(0)) \\ \downarrow & & \downarrow \\ GL((L(_1\alpha)/H_1)_0) & = & GL((L(_2\alpha)/H_2)_{\varphi(0)}) \end{array}$$

If $\Gamma_{i\alpha,H_i}$ denote the monodromy groups, it follows that $\Gamma_{1\alpha,H_1}$ is a subgroup of finite index in $\Gamma_{2\alpha,H_2}$.

Similarly, under the assumptions of 7.18, 7.19, 7.20 (i), the projective space bundle $B^{1''}$ descends to U and if $\Gamma_{i\mu,H_i}$ denotes the projective monodromy group, we have a commutative diagram

$$\begin{array}{ccc} \pi_1(\varphi^{-1}(U), 0) & \longrightarrow & \pi_1(U, \varphi(0)) \\ \downarrow & & \downarrow \\ \Gamma_{1\mu,H_1} & \longrightarrow & \Gamma_{2\mu,H_2} \end{array}$$

showing that $\Gamma_{1\mu,H_1}$ is a subgroup of finite index in $\Gamma_{2\mu,H_2}$.

Remark Cf. 8.33 for a slightly different definition of $\Gamma_{2\mu,H_2}$.

In §10, ψ will be the composite of $_1\tau : {}_1Q'' \to {}_1Q''/H_1$ and of an isomorphism φ. The inclusion above will hence be isomorphisms $\Gamma_{1\alpha,H_1} = \Gamma_{1\mu,H_1}, \Gamma_{1\mu,H_1} = \Gamma_{2\mu,H_2}$.

For φ an isomorphism, the ramification indices r_ψ are just the ramification indices for the map $_1\tau$.

7.22.

We have pointed out in 7.17 (i) (resp 7.20 (i) that the non integrality condition 7.15 (f) (resp. 7.18 (f')) can sometimes be dropped. Here we spell out when the cited continuity argument applies. Let $\begin{pmatrix} S \\ 2 \end{pmatrix}_f$ (resp. $\begin{pmatrix} S \\ 2 \end{pmatrix}_\infty$) be the set of pairs $\{u, v\} \in \begin{pmatrix} S \\ 2 \end{pmatrix}$ such that $_2\tau(D_{u,v}) = \psi(D)$ with D finite (resp. at infinity).

LEMMA 7.22.1. *If for all* $\{u, v\} \in \begin{pmatrix} S \\ 2 \end{pmatrix}_\infty$, *either*

(a) *the function* $x_u + x_v$ *on* $(\mathbf{C}^S)^{H_2}$ *is not a linear combination of* Σx_s *and of the* $x_s + x_t$ *for* $\{s, t\} \in \begin{pmatrix} S \\ 2 \end{pmatrix}_f$; *or*

(b) *it is such a linear combination. A value of* $1 - _2\mu_u - _2\mu_v$ *is thereby imposed by the conditions* $\Sigma\mu_s = 2$ *and (7.18) (e'): whenever* $\{s, t\} \in \begin{pmatrix} S \\ 2 \end{pmatrix}_f$ *with* $D_{s,t}$ *the image of a finite* $D \in {_1}I$,

$$1 - \ _2\mu_s - \ _2\mu_t = \frac{\tilde{r}_2(\psi(D))}{r_\psi(D)}.$$

It is assumed that for any $\{a, b\}$ *with* $\psi(D_{a,b}) = \ _2\tau(D_{u,v})$, *this imposed value is neither in* $\mathbf{Z}$ *nor in* $(\tilde{r}_2(D_{uv})/r_\psi(D_{ab}))\mathbf{Z}$,

then 7.20.1 applies: the condition 7.18 (f') = 7.15 (f) can be dropped in 7.19; it can be dropped as well as in 7.21 for the projective monodromy groups.

PROOF. Let $E \subset \mathbf{C}^S$ be the affine space of H_2-invariant S-uples μ with sum 2, and let E' be the affine subspace of μ for which $\Sigma\mu_s = 2$ and in addition: for all $\{s, t\} \in \begin{pmatrix} S \\ 2 \end{pmatrix}_f$, $\mu_s + \mu_t$ has the value given by 7.18 (e'). In order to apply the continuity argument asserted in 7.20 (i), we have to check that for $\{u, v\} \in \begin{pmatrix} S \\ 2 \end{pmatrix}_\infty$, the non-integrality conditions 7.15 (f) (= 7.18 (f')) hold generically on E'. These conditions are the ones assumed in (b) and have the form $A + B(\mu_u + \mu_v) \notin \mathbf{Z}$ with $B \neq 0$. Under hypothesis (a), $\mu_u + \mu_v$ is a non-constant linear function on E', so that $A + B(\mu_u + \mu_v)$ is generically non-integral. Under hypothesis (b), $\mu_u + \mu_v$ is constant on E', taking a value which is assumed to satisfy the non-integrality conditions. Thus 7.20.1 applies.

We now consider 7.17.1. let A^0 be the vector space of H_2-invariant families $\alpha_{st}(\{s, t\} \in \binom{S}{2})$ with the sum $\sum_t \alpha_{st} = 0$ for each s.

LEMMA 7.22.2. *If, in addition to the conditions of 7.22.1, the linear forms* α_{uv} *for* $\{u, v\} \in \binom{S}{2}_f$ $/H_2$ *are linearly independent on* A^0, *then 7.17 (i) applies: the condition 7.15 (f) can be dropped in 7.16; it can be dropped as well in 7.21.*

PROOF. For any H_2-invariant family α_{st} $(\{s, t\} \in \binom{S}{2})$, define a corresponding family $\mu_s = \sum_t \alpha_{st}$ with $\Sigma \mu_s = 2$. In 7.17 (i) we consider H_2-invariant families α satisfying 7.15 (a), (b), (c) and (e). For such a family α, condition 7.15 (c) is equivalent to condition 7.18 (e') on the corresponding μ. In order to apply 7.17 (i) we need only know that the μ corresponding to such an α generically satisfies the non-integrality condition 7.18 (f') $=$ 7.15 (f). The linear independence assumption implies that any H_2-invariant S-uple μ with sum 2 is obtained via $\mu_s = \Sigma \alpha_{st}$ from a H_2-invariant family $\alpha_{st}(\{s, t\} \in \binom{S}{2})$ with sum 1 and $\alpha_{uv} = 0$ for $\{u, v\} \in \binom{S}{2}_f$. This reduces 7.22.2 to 7.22.1.

§8. PRELIMINARIES ON MONODROMY GROUPS

8.1. Let P denote a complex projective line, N an integer ≥ 3, S a set of N elements, and M the subset of P^S consisting of injective maps from S to P. Let $\mu = (\mu_s)_{s \in S}$ be a family of real numbers with $0 < \mu_s < 1$ and $\sum_{s \in S} \mu_s = 2$; we call such a μ a *ball S-tuple*. Given a ball S-uple μ, a point $y \in P^S$ is called μ-stable (resp. semi-stable) if and only if for all $z \in P$,

$$\sum_{y(s)=z} \mu_s < 1 \ (resp. \ \leq 1).$$

The set of all μ-stable points (resp, μ-semi-stable points) is denoted

$$M_\mu^{st} \qquad (resp. \ M_\mu^{sst}).$$

Set

$$Q = M/Aut \ P, \qquad Q_\mu^{st} = M_\mu^{st}/Aut \ P,$$

where $Aut \ P(\simeq PGL(2, \mathbf{C}))$ acts diagonally on P^S. In DM, §4, a compactification Q_μ^{sst} of Q_μ^{st} is defined; it coincides with Q_μ^{st} if $M_\mu^{st} = M_\mu^{sst}$. The space Q_μ^{sst}. (resp. Q) is called the μ-moduli space of S-uples of points (resp. distinct points) of P. When the family μ is fixed in any discussion below, we shall write Q^{st} for Q_μ^{st}.

8.2. Let $\mu = \mu_{s \, s \in S}$ be a family of complex numbers such that $\mu \notin \mathbf{Z}$ and $\sum_{s \in S} \mu_s$ is an integer. Then for such an N-tuple ($N = |S|$) we have defined in [DM] (3.3) a subgroup Γ_μ of $PGL(N - 2, \mathbf{C})$.

In case $\mu_s \in \mathbf{R}$ for all $s \in S$, then the group Γ_μ, which depends only on $\mu \bmod \mathbf{Z}$, lies in a $PU(m, n)$ where

$$m = \left(\sum \ < \mu_s > \right) - 1, n = \left(\sum_{s \in S} \ < 1 - \mu_s > \right) - 1,$$

and where $< \mu_s >$ denotes the fractional part of μ_s, i.e., $0 \leq< \mu_s >< 1$,

$$\mu_s - < \mu_s > \in \mathbf{Z} \text{ for all } s \in S.$$

[DM] (2.21).

Let μ be a ball S-uple. We say μ satisfies condition (INT) if and only if for all $s, t \in S$ with $\mu_s + \mu_t < 1$ and $s \neq t$,

$$(1 - \mu_s - \mu_t)^{-1} \in \mathbf{Z}.$$

We say that μ satisfies condition $\Sigma INT(S_1)$ if and only if $S_1 \subset S, \mu_s = \mu_t$ for all $s, t \in S_1$ and for all $s, t \in S$ with $s \neq t$ and $\mu_s + \mu_t < 1$,

$$(1 - \mu_s - \mu_t)^{-1} \in \begin{cases} \frac{1}{2}\mathbf{Z} & \text{if } s, t \in S_1 \\ \mathbf{Z} & otherwise. \end{cases}$$

We say that μ satisfies condition ΣINT if μ satisfies $\Sigma INT(S_1)$ for some $S_1 \subset S$.

Let μ be a ball N-tuple with $N \geq 4$. In [DM] it is proved that Γ_μ is a lattice in $PU(1, N-3)$ if μ satisfies condition INT. In [M3] this hypothesis is weakened to condition ΣINT.

In [M3] it is proved that if $N - 3 > 1$ and μ is a ball N-tuple with Γ_μ discrete in $PU(N - 3, 1)$, then μ satisfies condition ΣINT except for nine ball 5-tuples and one ball 6-tuple. Some of these 10 Γ_μ, including the 6-tuple case, are arithmetic. The proof of discreteness in the remaining cases uses isomorphisms obtained by computer exploration in the dissertation of K. Sauter (cf. §9 below).

One of the principal objectives in this paper is to find a geometric proof of the commensurability of the remaining exceptional Γ_μ with Γ_ν where the ν satisfy condition ΣINT. Accordingly we shall deal mainly with ball 5-tuples.

8.3. In this section, we recall more explicit details from [DM] and [M3] about the group Γ_μ, adding related remarks.

Let P, S, M, Q, N be as in (8.1) and set

$$P_M = \{(p, m) \in P \times M; p \notin m(S)\}, \quad P_m = P - m(S).$$

Let μ be as in (8.2). On P_M there are (cf. [DM] (3.13)) flat one dimensional complex vector bundles L^μ with monodromy $= \exp 2\pi\sqrt{-1}\,\mu_s$ on each fiber P_m around the puncture $m(s)$ for $s \in S$. Then the

$$H^1(P_m, L_m^\mu)\ (m \in M)$$

are the fibers of a flat vector bundle over M, which induces a flat projective space bundle

(8.3.1) $B_Q^\mu \to Q$ (cf. [DM] (3.7))

which is uniquely determined by μ. The monodromy action of this flat bundle gives a homomorphism

$$\theta : \pi_1(Q, 0) \to Aut\ B_0^\mu \simeq PGL(N-2, \mathbf{C})$$

0 denoting a base point in Q. Set $\Gamma_\mu = Im\ \theta$. If the μ are real, the group Γ_μ lies in $PU(m, N-2-m)$ (cf. (8.2)). In case μ is a ball N-tuple, $\Gamma_\mu \subset PU(1, N-3)$.

The symmetric group $\Sigma(S)$ operates holomorphically on Q. For any $h \in \Sigma(S)$, let $F(h, Q)$ denote the fixed point set of h in Q.

LEMMA 8.3.2. $F(h, Q)$ is non-empty of complex codimension 1 in Q if and only if

(i) $N = 4$ and h is conjugate to (12), (123), or (1234) i.e., h is not in the Vierergruppe.

(ii) $N = 5$ and h is conjugate to $(12)(34)$

(iii) $N = 6$ and h is conjugate to $(12)(34)(56)$.

PROOF. If h is the identity, or if $N = 3$, or if $N = 4$ and h is in the Vierergruppe, $F(h, Q) = Q$. We may and shall exclude those cases.

Fix h and assume that $F(h, Q) \neq \phi$. Let $\tilde{F}$ be the inverse image of $F(h, Q)$ in M. It is the space of N-uples $(x_1, \ldots, x_N)$ for which there exists $g \in Aut\ P$ such that

(8.3.2.1) $(gx_1, \ldots, gx_N) = (x_{h^{-1}(1)}, \ldots, x_{h^{-1}(N)})$.

Since $N \geq 3$, the projectivity g is unique, of order the order m of h. Being of finite order m, it has two fixed points and, if z is a coordinate which has them for zero and pole, then, for a suitable m^{th} root of unity ζ, g is the map $z \to \zeta z$. It follows that h consists of cycles of length m plus at most two fixed points.

For g and h given as above, the dimension of the space of solutions of (8.3.2.1) is the number k of cycles of length m of h. Since $Aut\ P$ and h commute, the dimension of the space of solutions of (8.3.2.1) for some g is k plus the dimension of the conjugacy class of g: it is $k + 2$. The space $F(h, Q)$ is of codimension one in Q if and only if $\tilde{F}$ is of codimension one in M, and we are led to solve the equations

$$\begin{cases} N = km + a, & 0 \leq a \leq 2, m \geq 2 \\ N - 1 = k + 2, i.e., & k = N - 3 \end{cases}$$

The solutions (N, k, m, a) are $(4,1,2,2)$, $(4,1,3,1)$, $(4,1,4,0)$, $(5,2,2,1)$, $(6,3,2,0)$; 8.3.2 follows.

Remark Let Q^+ denote the subset of $Aut\ P \backslash P^S$ obtained by allowing two but no more than two coordinates of P^S to be equal, and let $D_{i,j}$ denote the divisor of Q^+ for which $x_i = x_j, i, j \in \binom{S}{2}$. Assume that $|S| \geq 5$. Reasoning as above we can show:

$$\text{If } D_{i,j} \subset F(h, Q^+), \text{ then } h = (ij).$$

Let H be a subgroup of $\Sigma(S)$. The N-tuple μ is H-invariant if $\mu_{h(s)} = \mu_s$ for $s \in S, h \in H$. Let $_H Q'$, or simply Q', denote the open subset of Q on which H acts freely; it is not empty if $|S| \geq 5$.

If μ is H-invariant, then the flat projective space bundle (8.3.1) descends to a flat projective space bundle

$$B^\mu_{Q'/H} \to Q'/H,$$

(c.f. [M3]2, where the case $H = \Sigma(S_1)$ is treated). Correspondingly, one gets the monodromy homomorphism

$$\theta_H : \pi_1(Q'/H, \bar{0}) \to Aut\ B^\mu_{\bar{0}} \approx PGL(N - 2, \mathbf{C})$$

where $\bar{0}$ is the image of the base point 0, which is taken in Q'.

Set

$$\Gamma_{\mu,H} = Im\ \theta_H.$$

Remark 8.3.3 Let Q'' be a Zariski open (dense) H-invariant subset of Q' with $\bar{0} \in Q''/H$, and let $i : \pi_1(Q''/H, \bar{0}) \to \pi_1(Q'/H, \bar{0})$. It is surjective; hence the image of $\pi_1(Q''/H, \bar{0})$ under $\theta_H \circ i$ is $\Gamma_{\mu,H}$. By abuse of notation, we shall denote $\theta_H \circ i$ by θ_H.

In 7.21 we have denoted by $\Gamma_{\mu,H}$ the projective monodromy group of an H-invariant local system V which is a twist of Lauricella hypergeometric functions associated to the H-invariant S-uple μ. The Lauricella system can be expressed locally in Q as (cf. (4.3), DM(3.5)):

$$x \to < H_1(P_0, \check{L}_0^\mu), \omega_\mu(x) > .$$

Thus as projective spaces with a Γ_μ action, $PV \approx PH_1(P_0, \check{L}_0^\mu)$, and as $\Gamma_{\mu,H}$ spaces also. The $\Gamma_{\mu,H}$ of 7.21 is defined as a subgroup of $Aut\ P(V/H)$. We can identify the latter with $Aut\ PV$ and the dual space of V with $H^1(P_0, L_0)$. Thus the $\Gamma_{\mu,H}$ of 7.21 is just the contragredient of the $\Gamma_{\mu,H}$ defined in this section.

In case μ is a ball N-tuple, the group $\Gamma_{\mu,H}$ preserves the positive ball $(B_+^\mu)_{\bar{0}}$ with respect to the positive cone defined by the invariant Hermitian form on $H^1(P_0, L_0^\mu)$ (DM(2.18)). Thus $\Gamma_{\mu,H} \subset Aut(B_+^\mu)_{\bar{0}} = PU(1, N-2)$.

8.3.4. Next we recall from [DM]3 the Schwartz map $\tilde{w}_\mu$ of $\tilde{Q}$ to the projective space $(B^\mu)_{\bar{0}}$. For each $m \in M$, there is, up to a factor, a unique element $w_\mu(m)$ in $H^1(P_m, L_m^\mu)$ described in [DM] (3.4): w_μ descends to a section of $(B^\mu)_Q$ over Q and of $(B_+^\mu)_{Q'/H}$ over Q'/H (cf. [M3] 2). Fixing a base point in the base of these flat fiberings, the section w_μ can be viewed as a multivalued map into the fiber over the base point and therefore as a single-valued map, the *Schwartz map*

$$\tilde{w}_\mu : \tilde{Q} \to (B^\mu)_0 \ (resp.\ \tilde{w}'_\mu : \widetilde{Q'/H} \to (B^\mu)_{\bar{0}})$$

where $\tilde{Q}$ (resp. $\widetilde{Q'/H}$) is the quotient of the universal covering of a Q (resp. Q'/H) by the kernel of monodromy homomorphism θ (resp. θ_H). The mapping $\tilde{w}_\mu : \tilde{Q} \to (B^\mu)_0$ is holomorphic and etale ([DM] (3.5) and (3.9)). Homogeneous projective coordinates for $\tilde{w}_\mu(x)$ can be obtained by integrating the form ω of 4.3 over a basis of cycles for $H^1(P_0, L_0^{dual})$; i.e., a basis of a Lauricella local system c.f. (4.4). By its definition, $\tilde{w}'_\mu$ is $\Gamma_{\mu,H}$-equivariant.

The covering map $Q' \to Q'/H$ yields

$$1 \to \pi_1(Q',0) \to \pi_1(Q'/H,\bar{0}) \to H \to 1.$$

We regard $\pi_1(Q',0)$ as a subgroup of $\pi_1(Q'/H,\bar{0})$. The kernel $Ker(\pi_1(Q',0) \to \pi_1(Q,0))$ lies in the kernel of θ_H by the etaleness of $\tilde{w}_\mu$. It follows at once that $\Gamma_\mu = \theta(\pi_1(Q,0)) = \theta_H(\pi_1(Q',0))$ and thus Γ_μ is a normal subgroup of $\Gamma_{\mu,H}$. The quotient $\Gamma_{\mu,H}/\Gamma_\mu$ is a quotient of H.

There are two cases in which $\Gamma_{\mu,H}/\Gamma_\mu$ can be computed easily.

CASE 1. $H = \Sigma(S_1)$, where $\mu_s = \frac{1}{2} - \frac{1}{p}, p$ odd for $s \in S_1$, and μ satisfies condition $\Sigma INT(S_1)$. By [M3] (3.11.1),

$$\theta_H(\pi_1(Q',0)) = \theta_H(\pi_1(Q'/\Sigma,\bar{0})).$$

Here $\Gamma_\mu = \Gamma_{\mu,\Sigma}$.

CASE 2. μ satisfies conditions INT and is H-invariant. Here $\Gamma_{\mu,H}/\Gamma_\mu = H$ by the same argument used in [M3] (3.11.2) in the case $H = \Sigma(S_1)$.

Let μ be a ball S-uple. Let $\widetilde{Q^{st}/H}$ denote the completion of the spread $Q'\tilde{/}H \to Q'/H$ over Q^{st}/H. In case $H = \Sigma(S_1)$ and μ satisfies condition $\Sigma INT(S_1)$, the completion of the map $w_\mu" Q'\tilde{/}H \to (B_+^\mu)_{\bar{0}}$ to $Q^{\tilde{s}t}/H$ yields a $\Gamma_{\mu,H}$-equivariant homeomorphism

(8.3.5) $$\tilde{w}_\mu : \widetilde{Q^{st}/H} \to (B_+^\mu)_{\bar{0}}$$

8.4. We define the 5-tuple $\mu = \mu(\pi,\rho,\sigma)$:

$$\mu(\pi,\rho,\sigma) = (\frac{1}{2} - \frac{1}{\pi}, \frac{1}{2} - \frac{1}{\pi}, \frac{1}{2} - \frac{1}{\pi}, \frac{1}{2} + \frac{1}{\pi} - \frac{1}{\rho}, \frac{1}{2} + \frac{1}{\pi} - \frac{1}{\sigma})$$

for which $\mu_1 = \mu_2 = \mu_3$. The condition $\Sigma\mu_i = 2$ is equivalent to

$$\frac{1}{\pi} + \frac{1}{\rho} + \frac{1}{\sigma} = \frac{1}{2}.$$

LEMMA 8.4.1.

(i) The 5-tuple (a,a,a,b,c) is a ball 5-tuple satisfying condition

$$\Sigma INT(S_1)$$

with $S_1 = \{1,2,3\}$ if and only if it is $\mu(\pi,\rho,\sigma)$ with $\frac{1}{\pi} + \frac{1}{\rho} + \frac{1}{\sigma} = \frac{1}{2}$ and

$$\pi, \frac{2\pi}{\pi - 6}, \rho, \sigma \in \mathbf{Z} \cup \infty, \pi \geq 3,$$

either $\rho \geq 2$ or $\rho < -6$,

and either $\sigma \geq 2$ or $\sigma < -6$.

(ii) The 5-tuple (a,a,a,a,c) is a ball 5-tuple satisfying condition

$$\Sigma INT(S_1)$$

*with $S_1 = \{1,2,3,4\}$ if and only if it is $\mu(\pi,\rho,\sigma)$ with $\frac{1}{\pi} + \frac{1}{\rho} + \frac{1}{\sigma} = \frac{1}{2}, \rho = \frac{\pi}{2},$
π integral ≥ 3, and $\frac{2\pi}{\pi - 6} \in \mathbf{Z} \cup \infty$.*

PROOF. Our ΣINT hypotheses on μ imply that $1 - \mu_4 - \mu_5 \leq 0$ or is the reciprocal of an integer n with $n \geq 2$. Consequently $\mu_4 + \mu_5 \geq \frac{1}{2}$ and $3a \leq \frac{3}{2}$. Condition ΣINT for μ_1, μ_2 yields $a = \frac{1}{2} - \frac{1}{\pi}$ with π an integer ≥ 3 or ∞.

In case (ii) $\pi = \infty$ is excluded, otherwise it would force $c = 0$. Our condition ΣINT implies that $3(\frac{1}{2} - \frac{1}{\pi}) - 1 = 1 - \mu_4 - \mu_5$, if positive, is the reciprocal of an integer. Hence $\frac{2\pi}{\pi - 6} \in \mathbf{Z} \cup \infty$ if it is positive.

One has $\frac{2\pi}{\pi - 6} < 0$ with π an integer ≥ 3 or ∞ only for $\pi = 3, 4, 5$. In each case $\frac{2\pi}{\pi - 6}$ is an integer. The proof of (ii) is now complete.

In case (i), we can assume, by symmetry, that $b \geq c$. Then $a + b \leq a + \frac{1}{2}(2 - 3a) = 1 - 2a < 1$. Therefore $1 - a - b$ is the reciprocal of an integer $\rho, \rho \in \mathbf{Z}, \rho \geq 2$. Set $\sigma^{-1} = 1 - a - c$. If $a + c < 1$ and $\sigma \in \mathbf{Z}, \sigma \geq 2$. If $a + c \geq 1$, then ΣINT imposes no further condition. The condition $\frac{1}{\pi} + \frac{1}{\rho} + \frac{1}{\sigma} = \frac{1}{2}$ then yields $\frac{1}{2} + \frac{1}{\pi} + \frac{1}{\rho} = 1 - \frac{1}{\sigma}$ with π, ρ as above. The only solutions for (pi, ρ, σ) are $(3,3,-6),(3,4,-12),(3,5,-30)$. Thus $\sigma \in \mathbf{Z}$. The solution $(3,3,-6)$ is excluded since it implies $c = 1$.

Remark The conditions $\pi, \frac{2\pi}{\pi-6} \in \mathbf{Z} \cup \infty, \pi \geq 3$ amount to $\pi \in \{3,4,5,6,7,8,9,10,12,18,\infty\}$. In case (ii), $\sigma = \frac{2\pi}{\pi - 6}$ can take as negative values only $-2,-4,-10$. In case (i) the only negative values σ can take are -12 and -30.

8.5. (cf. [DM] (4.5) Example 1). If $N = 5$ and $\mu_s + \mu_t < 1$ for all $s, t \in S, s \neq t$, then Q^{st} is compact and contains 10 lines of self intersection -1; they arise from the 10 codimension 1 diagonals in P^S. If the elements of S

are denoted A_1, A_2, A_3, B, C then we can picture the 10 lines schematically by the diagram

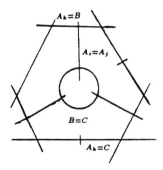

It will be convenient at times to weight the diagram by attaching the weight $1 - \mu_s - \mu_t$ to the line $s = t$.

This diagram has to be modified in case some $\mu_s + \mu_t \geq 1$. The lines with negative (resp. zero) weight have to be blown down to a point (resp. removed). If $\mu_s + \mu_t > 1$, the P^S diagonal $s = t$ is not contained in M^{st}; in its place is the codimension 2 diagonal $\{z \in P^S; z(s_1) = z(s_2)$ for $s_1, s_2 \in S - \{s, t\}\}$ whose image in Q_μ^{st} is a point.

8.6. Let H be a subgroup of $\Sigma(S)$. Let μ be a ball S-uple invariant under H. Let $_H Q'$ be the subset of Q on which H acts freely. For any $h \in H$ whose fixed points sets on Q has $\mathbf{C}$-codimension one, let D_h denote the set of $\mathbf{C}$-codimension one irreducible components of $F(h, Q)$. Let

$$\mathcal{D}_H = \bigcup \{D_h; dim_{\mathbf{C}} F(h, Q) = dim_{\mathbf{C}} Q - 1\}.$$

Let $\tau : Q^+ \rightarrow Q^+/H$ denote the projection. For any irreducible subvariety V of $\mathbf{C}$-codimension 1 of $Q^+/H -_H Q'/H$ let $[V]$ denote the conjugacy class in $\pi_1(_H Q'/H, 0)$ determined by a simple positive loop around V.

For any $s, t \in S$, set

$$e_{s,t} = \begin{cases} (1 - \mu_s - \mu_t)^{-1} & \text{if the transposition } (st) \text{ is not in } H \\ 2(1 - \mu_s - \mu_t)^{-1} & \text{if the transposition } (st) \text{ is in } H. \end{cases}$$

LEMMA 8.6.1. *Assume* $e_{s,t} \in \mathbf{Z}$ *for all* $s, t \in S$. *Let* N *denote the normal subgroup of* $\pi_1(_H Q'/H, \bar{0})$ *generated by*

$$\{[\tau(D)]^2; D \in \mathcal{D}_H\} \cup \{[\tau(D_{s,t})]^{e_{st}}; \{s, t\} \in \binom{S}{2}\}.$$

Then $N = Ker\ \theta_H$.

PROOF. Clearly $N \subset Ker\ \theta_H$. To prove the reverse inclusion, identify Q^+ and Q_μ^{st} with their corresponding subsets of $P^S/Aut\ P$ and set

$$Q_\mu^+ = Q_\mu^{st} \cap Q^+.$$

Set $Q' = {}_HQ'$ and let $\widehat{Q'/H}$ denote the simply connected covering space of Q'/H. Set ${}^*Q' = (\widehat{Q'/H})/N$, and let ${}^*Q^+$ denote the completion of the spread $*Q' \rightarrow Q'/H$ over Q_μ^+/H. It follows from (8.3.2) that $[\tau(D)]^2; D \in \mathcal{D}_H$ generates the kernel of $\pi_1({}_HQ',0) \rightarrow \pi_1(Q,0)$. By the universal mapping property of completions (cf. DM §8), there is a map $\eta :{}^* Q^+ \rightarrow Q_\mu^{st}/H$ (cf. definition preceding (8.3.5)); η is unramified over its image by etaleness over Q, and by matching ramification over $(Q_\mu^+ - Q)/H$.

The integrality assumption allows one to prove, just as in the case of hypothesis ΣINT, that $\tilde{w}_\mu : Q^{st}/H \rightarrow (B_+)_{\bar{0}}$ is a $\Gamma_{\mu,H}$ equivariant homeomorphism. Since $codim_{\mathbf{C}}(Q^{st} - Q^+) = 2$, we see that the complement of $Im\ \eta$ in $\widetilde{Q^{st}/H}$ is of $\mathbf{R}$-codimension 4. Inasmuch as the ball $(B_+)_{\bar{0}}$ is simply connected, so also is $Im\ \eta$. It follows at once that η is injective on ${}^*Q^+$. Inasmuch as $(\widehat{Q'/H})/N = (\widehat{Q'/H})/Ker\ \theta_H$, we get $N = Ker\ \theta_H$.

This implies that Q_μ^{st}/H is the biggest covering with the indicated ramifications over Q_μ^{st}/H.

§9. BACKGROUND HEURISTICS

In [M2, 1980], a group $\Gamma'(p,t)$ generated by three complex reflections of order p is associated to the Coxeter-like diagram

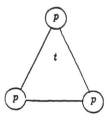

as follows:

To each node associate a base vector e_i and form the vector space $H : V = Ce_1 + Ce_2 + Ce_3$ with hermitian form Φ given by

$$< e_1, e_2 >=< e_2, e_3 > =< e_3, e_1 >= -\frac{1}{2 \sin \frac{\pi}{p}} \, e^{\pi\sqrt{-1}t/3}$$

$$< e_i, e_i >= 1;$$

define $R_i : V \rightarrow V$ via

$$v \mapsto v + (e^{2\pi\sqrt{-1}p} - 1) < v, e_i > e_i$$

($i = 1, 2, 3$). $\Gamma'(p,t)$ is defined to be the group generated by $\{R_1, R_2, R_3\}$. $\Gamma'(p,t)$ preserves the hermitian form Φ. The form Φ has signature (1 minus, 2 plus) for $|t| < 3(\frac{1}{2} - \frac{1}{p})$, ($p \neq 0, 1, 2$).

Subsequently in [M3], an explanation was given of the relation of $\Gamma'(p,t)$ to the monodromy groups Γ_μ of [DM], where μ is a ball 5-tuple with three equal μ_i. The two groups are related as follows. Let $\Gamma(p,t)$ denote the image of $\Gamma'(p,t)$ in $PU(1,2)$ and let C denote the cyclic group of order 3 induced by the cyclic permutation of the three generators of $\Gamma(p,t)$. Then

(9.1) $$C \, \Gamma(p,t) = \Gamma_{\mu,H}$$

where

$$\mu = (\frac{1}{2} - \frac{1}{p}, \frac{1}{2} - \frac{1}{p}, \frac{1}{2} - \frac{1}{p}, \frac{1}{4} + \frac{3}{2p} - \frac{t}{2}, \frac{1}{2} + \frac{3}{2p} + \frac{t}{2})$$

$$p = (\frac{1}{2} - \mu_1)^{-1}, \quad t = \mu_5 - \mu_4$$

$$H = \Sigma(\{1, 2, 3\}).$$

For each $(ij) \in \binom{S}{2}$, the monodromy of a positive interchange of punctures x_i and x_j for $i, j \in 1, 2, 3$ and a positive full turn of x_i around x_j otherwise, define Γ_μ conjugacy classes in $\Gamma_{\mu,H}$ of complex pseudo-reflections (these are of course complex reflections if they are of finite order), which are labelled

$$R(ij), A(i4), A'(i5), B'(45)$$

where $i, j \in 1, 2, 3$. In Γ_μ with μ satisfying condition INT, the 10 conjugacy classes $R^2(ij), A(i4), A'(i5), B'(45)(i, j \in 1, 2, 3)$ are distinct. The corresponding $\Gamma_{\mu,H}$ conjugacy classes of Γ_μ classes whose labels differ only by a permutation of $1, 2, 3$ coincide, thus yielding four classes of type R, A, A', B'.

Set

(9.2)
$$\frac{\pi}{2} = (1 - \mu_1 - \mu_2)^{-1}, \dot\rho = (1 - \mu_1 - \mu_4)^{-1}, \sigma = (1 - \mu_1 - \mu_5)^{-1},$$

$$\tau = (1 - \mu_4 - \mu_5)^{-1}.$$

From $\sum_i \mu_i = 2$, it follows that

(9.2)'
$$\pi^{-1} + \rho^{-1} + \sigma^{-1} = \frac{1}{2}.$$

μ satisfies condition $\Sigma INT(\{1, 2, 3\})$ if and only if π, ρ, σ, τ are integers, when positive. When any of them are integers, they represent the order of the cyclic groups of type R, A, A', B' respectively; these integers are positive, infinite, or negative according as the line of fixed points of the corresponding pseudo-reflection, in $\mathbf{P}^2$, intersects the open positive ball, is tangent to the boundary of the positive ball, or lies outside the positive ball respectively.

Three numbers π, ρ, σ satisfying (9.2)' determine a unique 5-tuple $\mu(\tau, \rho, \sigma)$ with $\mu_1 = \mu_2 = \mu_3$ and satisfying (9.1); that is,

$$(9.3) \qquad \mu(\pi, \rho, \sigma) = (\frac{1}{2} - \frac{1}{\pi}, \frac{1}{2} - \frac{1}{\pi}, \frac{1}{2} - \frac{1}{\pi}, \frac{1}{2} + \frac{1}{\pi} - \frac{1}{\rho}, \frac{1}{2} + \frac{1}{\pi} - \frac{1}{\sigma})$$

Set $H = \Sigma(1, 2, 3)$ and

$$(9.4) \qquad\qquad\qquad \Gamma(\pi, \rho, \sigma) = \Gamma_{\mu(\pi,\rho,\sigma),H}.$$

The search for a sufficient *and* necessary condition on a ball S-uple μ that the monodromy group Γ_μ be a lattice in $PU(1, |S| - 3)$ (cf. [M4]) led to the conjecture:

$\Gamma(\rho, 3, \sigma)$ can be embedded in $\Gamma(3, \rho, \sigma)$ for any integers ρ, σ such that $\frac{1}{\rho} + \frac{1}{\sigma} = \frac{1}{6}$.

In his doctoral dissertation ([S1]), Sauter proved this conjecture by exhibiting explicit embeddings

$$\Gamma(\rho, 3, \sigma) \rightarrow \Gamma(3, \rho, \sigma)$$

and in addition,

$$\Gamma(3, \rho, \sigma) \xrightarrow{onto} \Gamma(\sigma, 3, \rho), \ if \ 3 \mid \rho.$$

Sauter's method was based on a geometric study of the angles between the three fixed point sets of the complex reflections of type R in $\Gamma(\rho, 3, \sigma)$ and of type A in $\Gamma(3, \rho, \sigma)$ together with some computer exploration of angles between fixed point sets of triples of complex reflections in the conjugacy class of A. Finding the desired triples assured the conjugacy in the light of (9.1).

By a similar method, Sauter exhibited a conjugacy

$$(9.5) \qquad\qquad\qquad \theta : \Gamma_{1\mu, H_1} \hookrightarrow \Gamma_{2\mu, H_2}$$

where

$$_1\mu = \mu(\pi, 2, -\pi) = (\frac{1}{2} - \frac{1}{\pi}, \frac{1}{2} - \frac{1}{\pi}, \frac{1}{2} - \frac{1}{\pi}, \frac{1}{\pi}, \frac{1}{2} + \frac{2}{\pi})$$

$$_2\mu = \mu(\pi, \frac{\pi}{2}, \frac{2\pi}{\pi - 6}) = (\frac{1}{2} - \frac{1}{\pi}, \frac{1}{2} - \frac{1}{\pi}, \frac{1}{2} - \frac{1}{\pi}, \frac{1}{2} - \frac{1}{\pi}, \frac{1}{2} - \frac{4}{\pi})$$

$$(\frac{1}{2} - \frac{3}{\pi})^{-1} \in \mathbf{Z}, \ H_1 = \Sigma(\{1, 2, 3\}), \ H_2 = \Sigma(\{1, 2, 3, 4\}).$$

By comparing the volumes of $B/\Gamma_{1\mu,H_1}$ and $B/\Gamma_{2\mu,H_2}$ he could deter-mine that θ is an isomorphism. Inasmuch as $B/\Gamma_{\mu,H} = Q^{st}_\mu/H$ when μ satisfies condition $\Sigma INT(S_1)$ and $H = \Sigma(S_1)$, one concludes the remark-able isomorphism of spaces

$$(9.6) \qquad\qquad Q^{st}_{1\mu}/H_1 \simeq Q^{st}_{2\mu}/H_2.$$

The starting point of this paper was the effort to explain the non-obvious (9.6) directly and to deduce the isomorphism (9.5) from (9.6). The direct proof of (9.6) is given in §10 (cf. 10.15.4).

Remark The group $\Gamma'(p,t)$ can be defined for any $(p,t) \in \mathbf{C}^2$ with $p \neq 0$, replacing the hermitian form by the pairing of V with its dual. The equality $C\Gamma(p,t) = \Gamma_{\mu,\Sigma(3)}$ persists. Explicit formulae exhibiting these isomorphisms are found in Sauter's dissertation (cf. [Sauter, 1990]).

9.7. Let $_i\mu$ be a ball S-uple invariant under a subgroup H_i of the permutation group of $S(i = 1,2)$. Suppose we are given a morphism of analytic varieties

$$(9.7.1) \qquad\qquad \varphi : Q^+/H_1 \to \bar{Q}/H_2$$

where Q^+ is defined as above and $\bar{Q}$ is a partial compactification of Q. In these circumstances, we need a criterion for the Schwarz map of Q/H_1 (cf. (8.3.4)) to be the pullback of the Schwarz map of Q/H. We cannot expect the homogeneous projective coordinates of the Schwarz map given by hypergeometric functions (cf. (8.3.4), (4.4)) to be the pullbacks of one another. The theorems of §§6 and 7 will allow us to get the desired criterion by using twists of hypergeometric local systems (cf. Theorem 10.5).

§10. SOME COMMENSURABILITY THEOREMS

10.1. Let $(\mu_s)_{s \in S}$ be a family of numbers, $0 < \mu_s < 1$ with sum 2, and no partial sum equal to 1. The stable compactification Q_μ^{st} of Q (cf. 8.1) depends only on the set of subsets T of S with $\sum_{s \in T} \mu_s < 1$. If $N = |S| = 5$, it depends only on the set $\mathcal{T}$ of 2-elements subsets $\{s, t\}$ with $\mu_s + \mu_t > 1$. Indeed, for a 3-element subset T with complement $\{s, t\}$, one has $\sum_{u \in T} \mu_u < 1$ if and only if $\mu_s + \mu_t > 1$.

We fix $S = \{1, 2, 3, 4, 5\}$ and define $\overline{Q}_1$ to be the compactification corresponding to $\mathcal{T} = \phi$. We keep denoting by $D_{s,t}$ the divisor $x_s = x_t$ of $\overline{Q}_1$; these are the components of a divisor with normal crossing having complement Q. If one removes from $\overline{Q}_1$ the finitely many points $D_{s,t} \cap D_{u,v}$, one obtains Q^+. We define H_1 to be the subgroup of $\Sigma(S)$ generated by the transpositions $(1, 2)$ and $(3, 4)$. It is isomorphic to $\mathbf{Z}(2) \times \mathbf{Z}(2)$ and acts on $\overline{Q}_1$.

We define $\overline{Q}_2$ to be the compactification corresponding to $\mathcal{T} = \{1, 2\}$; the subgroup H_2 generated by the transposition $(3, 4)$ is isomorphic to $\mathbf{Z}(2)$ and acts on $\overline{Q}_2$.

The aim of the next three subsections is to construct an isomorphism

$$\varphi : \overline{Q}_1/H_1 \xrightarrow{\sim} \overline{Q}_2/H_2.$$

10.2. The space Q is the space of isomorphism classes of systems $(P; x_1, x_2, x_3, x_4, x_5)$ consisting of a projective line with five (ordered) distinct points on it. Similarly, the quotient $\overline{Q}_1/H_1$ is the space of isomorphism classes of systems $(P; A_1, A_2, B_1, B_2, C)$ with A_1, A_2 not ordered and B_1, B_2 not ordered, and with three of the marked points not allowed to coincide. The quotient $\overline{Q}_2/H_2$ is the space of isomorphism classes of systems $(P; A, B, F'_1, F'_2, C')$ with F'_1, F'_2

not ordered, $A \neq B$ and with three of the marked points not allowed to coincide, except for F_1', F_2', C'.

Let $D_{A,B}$ be the divisor on $\overline{Q}_1/H_1$ where some A_i equals some B_j. Similar notations will be used for the image in $\overline{Q}_i/H_i$ of divisors $D_{s,t}$. We first define φ on the complement of $D_{A,B}$ as an isomorphism

$$(10.2.1) \qquad \varphi : \overline{Q}_1/H_1 - D_{A,B} \xrightarrow{\sim} \overline{Q}_2/H_2 - D_{F_1',F_2'}.$$

Given $(P; A_1, A_2, B_1, B_2, C)$ with $A_i \neq B_j$, there is a unique non trivial involution I of P which interchanges A_1 and A_2 as well as B_1 and B_2. Let F_1, F_2 denote the fixed points of I. The quotient $P/ < I >$ is again a projective line. Let A, B, F_1', F_2', C' denote the image in $P/ < I >$ of $\{A_1, A_2\}, \{B_1, B_2\}$, the fixed points of I, and the C respectively. The map φ is defined by $(P; A_1, A_2, B_1, B_2, C) \mapsto (P/ < I >; A, B, F_1', F_2', C')$. The inverse is defined as follows: to $(P; A, B, F_1', F_2', C')$ one attaches the double covering $\tilde{P}$ of P ramified at F_1', F_2', the A_i above A, the B_i above B and a point above C': which one does not matter as the two systems $(\tilde{P}; A_1, A_2, B_1, B_2, C)$ obtained are isomorphic by a deck-transformation of $\tilde{P}/P$. The inverse of φ is

$$\varphi^{-1} : (P; A, B, F_1', F_2', C') \mapsto (\tilde{P}; A_1, A_2, B_1, B_2, C).$$

PROPOSITION 10.3. *The isomorphism φ extends to an isomorphism of $\overline{Q}_1/H_1$ with $\overline{Q}_2/H_2$. Divisors of $\overline{Q}_1/H_1$ are mapped to divisors of $\overline{Q}_2/H_2$ as follows:*

$A_i = C$	$B_i = C$	$A_1 = A_2$	$B_1 = B_2$	$A_i = B_j$	$C = F_i$
↓	↓	↓	↓	↓	↓
$A = C$	$B = C'$	$A = F_1'$	$B = F_1'$	$F_1' = F_2'$	$C' = F_i'$

PROOF. We offer two proofs, the first via calculations in coordinate systems, the second geometric and coordinate-free.

First proof Given any point $\overline{q}$ of $D_{A,B}$, choose $q \in D_{A_i,B_j}$ and choose a lift $m_0 \in P^S$ of q. Then $m_0(A_i) = m_0(B_j)$ and $m_0(S - A_i - B_j)$ consists of three distinct points. Assume at first that $\{A_1, A_2, B_1, B_2\}$ are three distinct points. For convenience we may assume $A_i = A_1, B_j = B_1$. Fix a coordinate z on P by thereby identifying P with $\mathbf{C} \cup \infty$. Let $\overline{M}_1$ denote the subset of points $m \in P^S$ such that $m(A_1) = 0$, $m(B_2) = 1$, $m(A_2) = \infty$,

$m(B_1) \neq \infty$. Without loss of generality, we can assume that $m_0 \in \overline{M}_1$. Since the *Aut P* orbit projection is injective on $\overline{M}_1$, we may identify $\overline{M}_1$ with a neighborhood U of q and take as coordinates on U:

$$(x,t) := (m(B_1), m(C)).$$

The involution I of P is $z \to \frac{x}{z}$. Choosing $z + \frac{x}{z}$ as a coordinate on P/I, the restriction of φ to U maps the isomorphism class with coordinates (x,t) to the isomorphism class of $(\mathbf{P}; \infty, x+1, 2\sqrt{x}, -2\sqrt{x}, t + \frac{x}{t})$ with $\sqrt{x}$ determined only up to sign. In order to see that this formula defines a morphism, set

$$\tilde{U} = \{(y,t), y^2 = x, t \in \mathbf{C} \cup \infty, x \neq \infty\},$$
$$\tilde{I} = \text{ involution } (y,t) \to (-y,t) \text{ on } \tilde{U}$$
$$\tilde{\varphi} : (y,t) \to (\mathbf{P}; \infty, y^2 + 1, 2y, -2y, t + \frac{x}{t}), \{2y, -2y\} \text{ unordered.}$$

Since $\tilde{\varphi}$ is a morphism and $\tilde{\varphi} \circ \tilde{I} = \tilde{\varphi}$, the map $\tilde{\varphi}$ descends to the morphism $\varphi : U \to \overline{Q}_2/H_2$, and it is indeed an isomorphism onto its image.

Suppose next that $\{A_1, A_2, B_1, B_2\}$ consists of only two points. We can assume for convenience that $q \in D_{A_1 B_1} \cap D_{A_2 B_2}$. To get coordinate systems in a neighborhood of $\bar{q}$ and in $\overline{Q}_2/H_2$, choose a coordinate z on the projective line P such that

$$z(m(A_1), m(A_2), m(B_1), m(B_2), m(C)) = (0, 1, x, 1-x, y).$$

I being the involution $z \to 1 - z$ with fixed points $\frac{1}{2}, \infty$. On P/I, take as coordinate $z' = z(1-z)$. Then

$$z'(m(A), m(B), m(F_1'), m(F_2'), m(C'))$$
$$= (0, x(x-1), \frac{1}{4}, \infty, y(1-y)).$$

Set $u = z'/z'(B)$. Inasmuch as A and B do not coalesce in $\overline{Q}_2/H_2$, we may take as coordinates on $\overline{Q}_2/H_2$

$$(v, w) := u(m(F_1'), m(C'))$$

and as coordinates on $\overline{Q}_1/H_1$ in a neighborhood of $\bar{q}$

$$(x, y) := z(m(B_1), m(C)).$$

The morphism φ is clearly holomorphic around $\bar{q}$, sending $\bar{q}$ to the point $m(F_1') = m(F_2') = m(C')$.

In particular, the morphism φ of 10.2 extends to a morphism $\overline{Q}_1/H_1 \to \overline{Q}_2/H_2$. Inasmuch as $\overline{Q}_1/H_1$ is compact, the image of φ is closed and hence is $\overline{Q}_2/H_2$. Since φ is an isomorphism on a Zariski-open neighborhood of each point of $\overline{Q}_1/H_1$ and is surjective, it is an isomorphism.

As B_1 approaches A_1, we have in the coordinates above $x \to 0$. Hence $\varphi(D_{A.B.}) = D_{F_1'F_2'} \subset \overline{Q}_2/H_2$.

The image under φ of the other divisors can be easily deduced from the description of φ in 10.2.

10.4. Next, we offer a geometric proof of Proposition 10.3. Let $_1\tau$ be the quotient map $\overline{Q}_1 \to \overline{Q}_1/H_1$. We will first construct a map from $\overline{Q}_1$ to $\overline{Q}_1/H_2$ which extends the composite map (10.2.1) $\circ_1\tau$. The space $\overline{Q}_2/H_2$ is a fine moduli space for systems consisting of a projective line P, points A, B, C' on P, and a degree two divisor $(F') = F_1' + F_2'$, with no three of A, B, C', F_1', F_2' coming together, except F_1', F_2', C', and with $A \neq B$. This means that, for any space X, to construct a map $u : X \to \overline{Q}_2/H_2$ amounts to constructing

(α) a projective line bundle $\pi : P \to X$;

(β) sections A, B, C' of π;

(γ) a relative divisor of degree two (F') with, fiber by fiber, the properties of non degeneracy listed above holding.

We take $X = \overline{Q}_1$ and proceed to construct such a system $(\alpha)(\beta)(\gamma)$, extending the one which defines the map (10.2.1) $\circ_1\tau$ on $\overline{Q}_1 - \cup D_{A_i,B_j}$.

Let $P_{\overline{Q}_1} \to \overline{Q}_1$ be the universal family of projective lines with 5 marked points, parametrized by $\overline{Q}_1$. It admits the 5 sections A_1, A_2, B_1, B_2, C. For $i, j \in \{1, 2\}$, the sections A_i and B_j are smooth divisors in $P_{\overline{Q}_1}$, intersecting transversely. Let $E_{i,j}$ be their intersection, mapping via $_1\tau$ isomorphically to D_{A_i,B_j}. Let $\tilde{P}$ be obtained by blowing up the $E_{i,j}$. The fiber of $\tilde{P}$ above $x \in \overline{Q}_1$ coincides with that of $P_{\overline{Q}_1}$ if x is in no D_{A_i,B_j}, consists of two projective lines joined at one point if x is in one D_{A_i,B_j} and consists of three linked projective lines if x is $D_{A_1,B_1} \cap D_{A_2,B_2}$ or $D_{A_1,B_2} \cap D_{A_2,B_1}$. Here is the picture of what happens above a transversal slice at a point of D_{A_1,B_1} not in D_{A_2,B_2}.

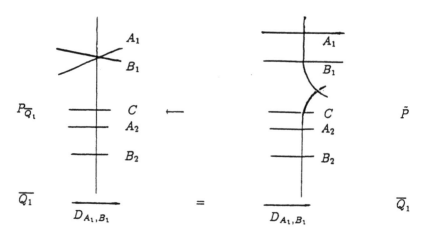

A similar picture is valid above D_{A_i,B_j} and, at $D_{A_1,B_1} \cap D_{A_2,B_2}$, the special fiber is

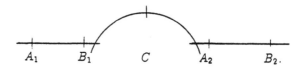

The morphism $\pi : \tilde{P} \to \overline{Q}_1$ is a flat local complete intersection morphism. The relative dualizing sheaf ω is hence a line bundle, with restriction to any fiber F the dualizing sheaf of F. For each fiber, one checks that

$$H^1(F, \omega^{-1}) = 0,$$

that ω^{-1} is generated by its global sections, that

$$\dim H^0(F, \omega^{-1}) = 3,$$

and that the map from F to a projective plane defined by ω^{-1} maps F to a conic. For each of the three kinds of fibers, we give a description of the restriction of ω^{-1} to each irreducible component and of the map to $\mathbf{P}^2$ defined by a basis of $H^0(F, \omega^{-1})$:

$\overline{\mathcal{O}(2)}$:embedding as a non singular conic

$A_i = B_j$: $\mathcal{O}(1) \quad \mathcal{O}(1)$:embedding as a conic degenerated into two projective lines

$A_1 = B_1$: $\mathcal{O}(1) \quad \mathcal{O} \quad \mathcal{O}(1)$:the middle component is
$A_2 = B_2$:   contracted to a point. The image is again a degenerate conic.

Because of the H^1 vanishing, the sheaf $\pi_*\omega^{-1}$ is a vector bundle and its fiber at x maps isomorphically to $H^0(\pi^{-1}(x), \omega^{-1})$ (EGA III 7.8.4 b) $\Rightarrow$ d) and 7.7.4, 7.5.5(11)). Let $\mathbf{P}(\pi_*\omega^{-1})$ be the projective plane bundle defined by the vector bundle $\pi_*\omega^{-1}$. We use here Grothendieck's definition: sections of $\mathbf{P}(\pi_*\omega^{-1})$ correspond to quotient line bundles of $\pi_*\omega^{-1}$. We have a natural map

$$f : \tilde{P} \to \mathbf{P}(\pi_*\omega^{-1}).$$

A local basis (e_1, e_2, e_3) of $\pi_*\omega^{-1}$ identifies (locally on $\overline{Q}_1$) $\mathbf{P}(\pi_*\omega^{-1})$ with $\mathbf{P}^2 \times \overline{Q}_1$, and (e_1, e_2, e_3) are projective coordinates for f. The image of f is a divisor $f(\tilde{P})$ of $\mathbf{P}(\pi_*\omega^{-1})$. The fiber by fiber description given above shows that $f(\tilde{P})$ is a family of plane conics, parametrized by $\overline{Q}_1$.

For $x \in \overline{Q}_1$, x not on any D_{A_i, B_j}, the involution I of the fiber P_x, the divisor of fixed points, and the quotient map $P_x \to P_x/ < I >$ can be described as follows. One has $P_x \subset \mathbf{P}^2 = \mathbf{P}(H^0(P_x, \omega^{-1}))$. Let a (resp. b) be the line spanned by A_1, A_2 (resp. B_1, B_2). It becomes the tangent to the conic $P_x \subset \mathbf{P}^2$ when the points coalesce. Let O be the intersection point of a and b. The involution I assigns to each point z the other intersection point of the line Oz with the conic P_x.

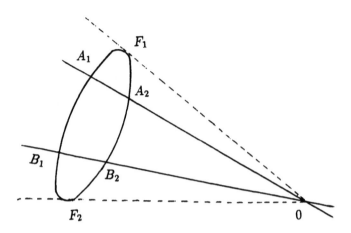

The quotient $P_x/ < I >$ is the projective line parametrizing the lines through O, and the quotient map is

$$x \in P_x \mapsto \text{line } Ox.$$

The divisor of fixed points is the intersection of the conic P_x with the polar line of O.

Over $\overline{Q}_1$, we get a family of projective planes $\mathbf{P}H^0(\tilde{P}_x, \omega^{-1})$ parametrized by $\overline{Q}_1$, with total space $\mathbf{P}(\pi_*\omega^{-1})$, and on them two distinct lines a and b intersecting in O.

We can view O as a section of $\mathbf{P}(\pi_*\omega^{-1})$ over $\overline{Q}_1$. Let P_2 be the projective line bundle on $\overline{Q}_1$ whose fiber at $x \in \overline{Q}_1$ is the projective line of lines through O. Taking for $z \in \tilde{P}$ the line through $f(z)$ and O we obtain a $\overline{Q}_1$-map g from $\tilde{P}$ to P_2. We define the section A of $P_2/\overline{Q}_1$ as being $g(A_1) = g(A_2)$, B as $g(B_1) = g(B_2)$, C' as $g(C)$ and the degree two divisor (F') as the image of the degree two divisor on $f(\tilde{P})$, intersection of $f(\tilde{P})$ with the polar line of O. At a point of D_{A_1,B_1} (resp. $D_{A_1,B_1} \cap D_{A_2,B_2}$), the picture is as follows:

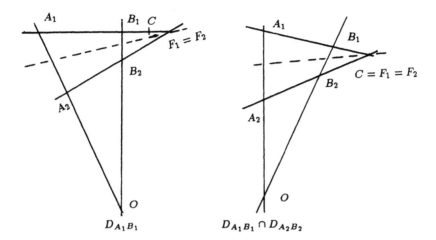

$$D_{A_1 B_1} \qquad\qquad D_{A_1 B_1} \cap D_{A_2 B_2}$$

(the dotted line is the polar of O).

The system $(P_2; A, B, C', (F'))$ gives the required map from $\overline{Q}_1$ to $\overline{Q}_2/H_2$. Over Q_1, hence everywhere, it is H_1-invariant. It factors through

$$\overline{\varphi} : \overline{Q}_1/H_1 \to \overline{Q}_2/H_2.$$

The map $\overline{\varphi}$ is bijective. Indeed, tracing back the construction we gave, one sees that it has as inverse the map attaching to the isomorphism class of $(P; A, B, F_1', F_2', C')$ the isomorphism class of the following projective line with marked points:

Case (a) $F_1' \neq F_2'$: One considers the double covering of P ramified at F_1' and F_2' and on it the inverse images A_1, A_2 of A, B_1, B_2 of B, and one inverse image of C'.

Case (b) $F_1' = F_2' \neq C'$: one considers two copies of P, glued together at $F_1' = F_2'$. One takes the inverse images A_1, A_2 of A, B_1, B_2 of B, one inverse image C of C' and one then contracts to a point the component not containing C. This amounts to taking P with the points $A_1 = A, A_2 = F_1' = F_2'$, $B_1 = B, B_2 = F_1' = F_2'$, $C = C'$.

Case (c) $F_1' = F_2' = C'$: one takes $A_1 = B_1, A_2 = B_2$.

Being bijective, the map $\overline{\varphi}$ is an isomorphism. The correspondence between divisors of $\overline{Q}_1/H_1$ and $\overline{Q}_2/H_2$ is left to the reader. It follows from the geometric description of $\overline{\varphi}$ which has been given. The map $\overline{\varphi}$ coincides with the φ of 10.3.

10.5. We now apply 7.19 and 7.22.1 to the map $\psi = \varphi \circ {}_1\tau$. More precisely, we choose ${}_iQ''$ to be the complement of finitely many H_i-orbits in ${}_i\overline{Q}$ in such a way, that ${}_iQ'' \subset Q^+$, that ${}_iQ''/H_i$ is smooth, that ${}_iQ''/H_i - {}_iQ'/H_i$ is a disjoint union of smooth divisors ($i = 1, 2$), and that ψ induces an isomorphism from ${}_1Q''/H_1$ to ${}_2Q''/H_2$. The exact choice of ${}_iQ''$ is irrelevant, cf. 7.15. We apply 7.19 to the map ψ from ${}_1Q''$ to ${}_2Q''/H_2$.

The ramification divisors for the quotient maps ${}_i\tau$: ${}_i\overline{Q} \rightarrow {}_i\overline{Q}/H_i$ are the divisors of fixed points of elements of H_i. The elements of H_1 acting on $\overline{Q}_1$ have the following divisors of fixed points (they may have additional isolated fixed points):

$$(12) \in H_1 : D_{A_1,A_2}$$
$$(34) \in H_1 : D_{B_1,B_2}$$
$$(12)(34) \in H_1 : D_{F,C} \qquad (cf.8.3.2)$$

The map ${}_1\tau$ is ramified with ramification index 2 along these divisors. The map φ being an isomorphism by 10.3, ψ has the same ramification as ${}_1\tau$. For H_2 acting on $\overline{Q}_2$: $(3,4) \in H_2$ has D_{F_1,F_2} as divisor of fixed points and ${}_2\tau$ ramifies with ramification index 2 along this divisor.

With the notations of 7.15, ${}_1I$ consists of all divisors at infinity $D_{s,t}$, and of the divisor at finite distance D_{FC}, while ${}_2I$ consists only of divisors at infinity: the images of $D_{s,t}$, with D_{AB} excluded. The assumptions (a)(b)(c) of 7.15 are satisfied.

As in 7.19, let ${}_2\mu$ be an H_2-invariant 5-uple of complex numbers with sum 2. Let us use the symbols ABFFC for $1, 2, 3, 4, 5$ to build the H_2-invariance into the notation. Similarly, for ${}_1\mu$, we will use AABBC for $1, 2, 3, 4, 5$. We assume the condition (e') of 7.18; it amounts to

$$(10.5.1) \qquad\qquad {}_2\mu_C + {}_2\mu_F = \frac{1}{2}.$$

The hypothesis (a) of 7.22.1 holds; that is, on the space of 4-uples $(\mu_A, \mu_B, \mu_F, \mu_C)$ none of the linear forms

$$\mu_A + \mu_F, \mu_B + \mu_F, \mu_A + \mu_C, \mu_B + \mu_C, 2\mu_F$$

is a linear combination of the linear forms $\mu_A + \mu_B + 2\mu_F + \mu_C$ and $\mu_C + \mu_F$. By 7.19 and 7.22.1, we get that if $_1\mu$ is defined by

(10.5.2)
$$1 - 2 \,_1\mu_A = 2(1 - \,_2\mu_A - \,_2\mu_F)$$
$$1 - 2 \,_1\mu_B = 2(1 - \,_2\mu_B - \,_2\mu_F)$$
$$1 - \,_1\mu_A - \,_1\mu_B = \frac{1}{2}(1 - 2 \,_2\mu_F)$$
$$1 - \,_1\mu_A - \,_1\mu_C = 1 - \,_2\mu_A - \,_2\mu_C$$
$$1 - \,_1\mu_B - \,_1\mu_C = 1 - \,_2\mu_A - \,_2\mu_C$$

the flat projective space bundle $B^{2''}/H_2$ on $_2Q'/H_2$ corresponds by φ to the flat projective space bundle $B^{1''}/H_1$ on $_1Q'/H_1$.

The affine space of allowed $_2\mu$ (sum 2, H_2-symmetric, satisfying (10.5.1)) maps by (10.5.2): $_2\mu \mapsto \,_1\mu$ to the affine space of H_1-symmetric, sum 2, 5-uples. The map is injective and both spaces are of dimension 2. It is an isomorphism, with corresponding $_i\mu$ given by $_1\mu = (a, a, b, b, 2 - 2a - 2b)$, $_2\mu = (1 - b, 1 - a, a + b - \frac{1}{2}, a + b - \frac{1}{2}, 1 - a - b)$. We conclude by 7.21 and 7.22.1:

THEOREM 10.6. *Let*

(10.6.1)
$$_1\mu = (a, a, b, b, 2 - 2a - 2b),$$
$$H_1 = < (1, 2), (3, 4) >$$
$$_2\mu = \left(1 - b, 1 - a, a + b - \frac{1}{2}, a + b - \frac{1}{2}, 1 - a - b\right)$$
$$H_2 = < (3, 4) >$$

Then $\Gamma_{_1\mu H_1}$ is conjugate in $PGL(3, \mathbf{C})$ to $\Gamma_{_2\mu H_2}$.

COROLLARY 10.7. *Suppose that $_1\mu$ and $_2\mu$ are as above. Assume that $0 < a < 1, 0 < b < 1, 0 < b < 1$, and $\frac{1}{2} < a + b < 1$; or equivalently that $0 < \,_1\mu_i < 1$ (resp. $0 < \,_2\mu_i < 1$). Then, the birational map φ of 10.3 extends to an isomorphism*

$$Q^{sst}_{_1\mu}/H_1 \xrightarrow{\sim} Q^{sst}_{_2\mu}/H_2.$$

PROOF. $Q^{sst}_{_i\mu}$ is obtained from $\overline{Q}_1$ by contracting the divisors $D_{j,k}$ for which $_i\mu_j +_i \mu_k \geq 1$. For $Q^{st}_{_2\mu}$, the divisor D_{AB} is always to be contracted:

$_2\mu_5 > 0$ amounts to $_2\mu_1 + _2\mu_2 > 1$. Each $Q^{sst}_{2\mu}$ is hence a contraction of $\overline{Q}_2$. For corresponding divisors, corresponding $1 - \mu_\bullet - \mu_\bullet$ are positive multiples of one another, hence corresponding divisors are contracted simultaneously. The corollary follows.

10.8. We will need the following particular case.

The compactification $\overline{Q}_2$ of Q corresponding to T consisting of $\{1, 2\}$ is deduced from $\overline{Q}_1$ by contracting $D_{A_1 A_2}$ to a point. Let $\overline{Q}_3$ be the compactification of Q corresponding to $\{\{1, 2\}, \{1, 3\}, \{1, 4\}\}$; it is deduced from $\overline{Q}_2$ by contracting each of the disjoint divisors $D_{AF'_1, AF'_2}$ to a point. Since $D_{AA} \subset \overline{Q}_1 / H_1$ corresponds to $D_{AF'} \subset \overline{Q}_2 / H_2$, we get that

COROLLARY 10.9. *With the notations above, the birational map φ extends to an isomorphism*

$$\overline{Q}_2 / H_1 \xrightarrow{\sim} \overline{Q}_3 / H_2.$$

10.10. Suppose in (10.6.1) $a = b$. Then

$$_1\mu = (a, a, a, a, 2 - 4a), \quad _2\mu = (1 - a, 1 - a, 2a - \frac{1}{2}, 2a - \frac{1}{2}, 1 - 2a).$$

We can then apply (10.6) to $_2\mu$ and

$$_3\mu = (\frac{3}{2} - 2a, a, a, a, \frac{1}{2} - a)$$

obtaining th conjugacy of $\Gamma_{_2\mu H'_2}$ and $\Gamma_{_3\mu H_3}$ where

$$H'_2 = < (12), (34) >, \quad \text{and } H_3 = < (3, 4) > .$$

The composition of these two results gives:

$$\Gamma_{_1\mu, H_1} \text{ is conjugate to a subgroup of finite index in } \Gamma_{_3\mu, H_3}.$$

However, we can relate these groups more directly.

10.11 Set $S_1 = \{1, 2, 3, 4\}$ and let $V(S_1) \subset \Sigma(S)$ denote the Vierergruppe of S_1. We recall that it is a normal subgrouop of $\Sigma(S_1)$, that its three nontrivial elements are the fixed-point-free involutions of S_1 and that, acting by inner conjugation on $V(S_1), \Sigma(S_1)$ maps into the full symmetric group of $V(S_1) - \{e\}$ with kernel $V(S_1)$. We choose a labelling I_2, I_3, I_4 of the three non trivial elements of $V(S_1)$, hence a morphism

(10.11.1) $\Sigma(S_1) \to \Sigma(\{2,3,4\}) \subset \Sigma(S)$.

The symmetric group $\Sigma(S_1)$ acts on $\overline{Q}_1/V(S_1)$ via its quotient $\Sigma(\{2,3,4\})$. The symmetric group $\Sigma(\{2,3,4\})$ also acts on the compactification $\overline{Q}_3$ of Q corresponding, as in 10.1, to $\mathcal{T} = \{\{1,2\},\{1,3\},\{1,4\}\}$. We will construct a $\Sigma(\{2,3,4\})$ equivariant isomorphism

(10.11.2) $\psi : \overline{Q}_1/V(S_1) \xrightarrow{\sim} \overline{Q}_3$.

Let $\overline{Q}_1' \subset \overline{Q}_1$ be the open set which is the moduli space of systems consisting of a projective line P with 5 points (A_1, A_2, A_3, A_4, C) on it, with $A_i \neq A_j$ for $i \neq j$. The quotient $\overline{Q}_1'/V(S_1)$ is then the moduli space of isomorphism classes of systems consisting of a projective line P, an unordered set of four distinct points $\{A_1, A_2, A_3, A_4\} \subset P$, a fifth point C, and a V-structure on $\{A_1, A_2, A_3, A_4\}$, that is a labelling I_2, I_3, I_4 of the three fixed-point-free involutions of $\{A_1, A_2, A_3, A_4\}$.

Each of the three fixed-point-free involutions I of $\{A_1, A_2, A_3, A_4\}$ extends to an involution of P, denoted by the same letter and this construction defines an action of the Vierergruppe V on P., The quotient P/V is again a projective line. For $i = 2,3,4$, the two fixed points of the involution I_i are permuted by each of the other I_j. Let F_i be the image of the fixed points of I_i in P/V. Let A'' be the image of the A's, and C'' be the image of C. The restriction ψ_0 of ψ to $\overline{Q}_1'/V(S_1)$ is defined as:

$$(P; A_1, A_2, A_3, A_4, C; I_2, I_3, I_4) \mapsto (P/V; A'', F_2, F_3, F_4, C'').$$

It maps $\overline{Q}_1'/V(S_1)$ isomorphically to the open subset of $\overline{Q}_3$ parametrizing systems $(P; A'', F_2, F_3, F_4, C'')$ with the F_i and A all distinct. The inverse of ψ is defined by assigning to $(P; A'', F_2, F_3, F_4, C'')$ the composition $\tilde{P}$ of the double coverings of P ramified at two of the F's, the unordered set A of points of $\tilde{P}$ above A'', any point C of $\tilde{P}$ above C'', and the labelling I_2, I_3, I_4 of the fixed point free involutions of A, where I_i extends to the involution of $\tilde{P}$ whose fixed points map to F_i; it is the one such that $\tilde{P}/< I_i >$ is the double covering of P ramified at the $F_j, j \neq i$.

To conclude the construction of ψ, it remains to prove:

PROPOSITION 10.12. *(i) The map $\psi_0 : \overline{Q}'_1/V(S_1) \to \overline{Q}_3$ constructed above extends to an isomorphism ψ of $\overline{Q}_1/V(S_1)$ with $\overline{Q}_3$.*
(ii) Divisors D_{ij} correspond as follows:

(a) *$A_i = C$ maps to $A'' = C''$*

(b) *($C = $ a fixed point of I_i) maps to $C'' = F_i$*

(c) *for I_k denoting the involution $(a,b)(c,d)$ of S_1 and I_ℓ, I_m the two others,*

$$(A_a = A_b \text{ or } A_c = A_d) \text{ maps to } F_\ell = F_m.$$

Let $\sigma \in \Sigma(\{2,3,4\})$ be a transposition. Let $H \subset \Sigma(S_1)$ be the inverse image in $\Sigma(S_1)$ of the subgroup $<\sigma>$ of $\Sigma(\{2,3,4\})$. Dividing by $<\sigma>$, we deduce from

(10.12.1) $$\psi_0 : \overline{Q}'_1/V(S_1) \to \overline{Q}_3$$

a map

(10.12.2) $$\psi_1 : \overline{Q}'_1/H \to \overline{Q}_3/<\sigma>$$

LEMMA 10.13. *The map (10.12.1) extends to an isomorphism ψ_H of $\overline{Q}_1/H$ with $\overline{Q}_3/<\sigma>$.*

PROOF. We may and shall assume that the I_i are chosen as follows: $I_2 : (12)(34)$, $I_3 : (13)(24)$, $I_4 : (14)(23)$, and that $H \subset \Sigma(S_1) \subset \Sigma(S)$ is the subgroup of $\Sigma(S_1)$ respecting the partition $(12)(34)$. This corresponds to $\sigma = (3,4)$.

The isomorphism (10.3)

$$\overline{Q}_1/(\mathbf{Z}/(2) \times \mathbf{Z}/(2)) \to \overline{Q}_2/(\mathbf{Z}/(2))$$

is $\mathbf{Z}/(2)$-equivariant, for the action permuting the A's and B's at left, and A and B at right. Passing to the quotient, we get an isomorphism

(10.13.1) $$\overline{Q}_1/H \xrightarrow{\sim} \overline{Q}_2/(\mathbf{Z}/(2) \times \mathbf{Z}/(2))$$

with the group at the right generated by $(1,2)$ (permute A and B) and $(3,4)$ (permute F_1 and F_2). It is H_1. Let us apply 10.8 to $\overline{Q}_2/\mathbf{Z}/(2) \times \mathbf{Z}/(2)$. We obtain an isomorphism

(10.13.2) $$\overline{Q}_2/\mathbf{Z}/(2) \times \mathbf{Z}/(2) \xrightarrow{\sim} \overline{Q}_3/H_2,$$

with H_2 generated by the transposition $(3,4)$.

Let us check that on $\overline{Q}'_1/H$, the composite of the isomorphism (10.13.1) and (10.13.2) agrees with ψ_1. The image of the point corresponding to $(P; A_1, A_2, A_3, A_4, C)$ is obtained as follows:

(a) Let F_1 and F_2 be the fixed points of the involution $I_2 = (A_1 A_2)(A_3 A_4)$ of P, $\pi : P \to P/ < I_2 >$ be the quotient map, and take $A = \pi(A_1) = \pi(A_2)$, $B = \pi(A_3) = \pi(A_4)$, $F'_i = \pi(F_i)$, $C' = \pi(C)$. This gives us five points A, B, F'_1, F'_2, C' on $P/ < I_2 >$.

(b) Let τ be the involution of $P/ < I_2 >$ exchanging A and B, as well as F'_1 and F'_2. It is the quotient by $< I_2 >$ of the involution $I_3 = (A_1, A_3)(A_2, A_4)$ of P, which commutes with I_2. One has

$$< I_3, I_2 >= \text{Vierergruppe } V \text{ of } \{A_1, A_2, A_3, A_4\}$$

and $(P/ < I_2 >)/ < I_3 >= P/V$.

The points to be taken on P/V are the image A of A_i, the image of the fixed point of I_2, the images of the fixed points of τ, and the image of C. One of the two fixed points of τ is the image in $P/ < I_2 >$ of the fixed points of I_3, and the other image in $P/ < I_2 >$ of the fixed points of I_4. We have recovered the description (10.11) of ψ. Composing (10.13.1) and (10.13.2) thus yields (10.13).

10.14. Proof of 10.12: It follows from 10.13 that the closure of the graph of ψ_0 is a finite to finite correspondence from $\overline{Q}_1/V(S_1)$ to $\overline{Q}_3$. It is also generically one to one. The source and target being normal varieties, it is an isomrophism ψ.

Alternatively, the space $\overline{Q}_1/V(S_1)$ is the completion (cf. [DM] (8.1)) of $\overline{Q}'_1/V(S_1)$ over $\overline{Q}_1/H$, and by its universal property, there is a continuous mapping $\psi : \overline{Q}/V(S_1)$ to $\overline{Q}_3$ so as to make the diagram below commutative

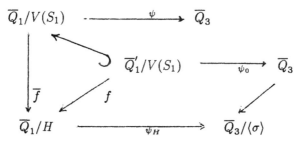

Thus ψ yields the isomorphism.

For the map 10.13, the effect of ψ_H on interesting divisors is obtained by applying 10.3 twice: $D_{A,C}$ is mapped to $D_{A'',C''}$, the divisor $A_1 = A_2$ or $A_3 = A_4$ is mapped to $F_3 = F_4$, and $A_i = A_j (i = 1, 2, j = 3, 4)$ to $F_3 = F_4$. The fixed point set of $I_2 = (12)(34)$ is mapped to $C'' = F_2$.

For ψ, it follows that $D_{A,C}$ maps to $D_{A'',C''}$, that $A_1 = A_2$ or $A_3 = A_2$ or $A_3 = A_4$ maps to $F_3 = F_4$, and that the fixed point set of I_2 maps to $C'' = F_2$. This, and $\Sigma(S_1)$-equivariance of ψ, forces 10.12.

Remark 10.15 The proof 10.13 showed that the following commutative diagram relates the isomorphism ψ of 10.12, the isomorphism 10.3, and the isomorphism 10.9 deduced from it by contraction. Vertical arrows are quotient maps and $H = H_1 V$.

(10.15.1)

$$
\begin{array}{ccc}
\overline{Q}_1/V & \xrightarrow{\quad 10.12 \quad} & \overline{Q}_3 \\
\downarrow & & \downarrow \\
\overline{Q}_1/H \xrightarrow{10.13} \overline{Q}_2/H_1 \xrightarrow{10.9} & & \overline{Q}_3/H_2 \\
\uparrow & & \uparrow \\
\overline{Q}_1/H_1 & \xrightarrow{10.13} & \overline{Q}_2/H_2
\end{array}
$$

Setting $\Sigma_4 = \Sigma(S_1)$ and $\Sigma_3 = \Sigma(\{F_2, F_3, F_4\})$, we have $\Sigma_4 = \Sigma_3 V$, from which we infer at once the conjugacies in $PGL(3, \mathbf{C})$:

(10.15.2) $\Gamma_{1\mu,V} \xrightarrow{\sim} \Gamma_{3\mu}, \ \Gamma_{1\mu,\Sigma_4} \xrightarrow{\sim} \Gamma_{3\mu,\Sigma_3}.$

Reasoning as in the proof of 10.7, for $\frac{1}{2} < a < 1$ or equivalently, for $0 < {}_1\mu_j < 1$ (resp. $0 < {}_3\mu_j < 1$) we get that the isomorphism ψ of 10.12 yields isomorphisms

(10.15.3) $Q_{1\mu}^{sst}/V(S_1) \xrightarrow{\sim} Q_{3\mu}^{sst}$

(10.15.4) $Q_{1\mu}^{sst}/\Sigma_4 \xrightarrow{\sim} Q_{3\mu}^{sst}/\Sigma_3$

As pointed out in our introduction, this last isomorphism was discovered by K. Sauter in his thesis and touched off our investigation.

10.16. We now add to the hypotheses of 10.6 the assumption that we are dealing with ball S-uples μ i.e.,

$$0 < \mu_s < 1 \text{ for } s \in S.$$

In [DM] (3.10) there is defined a hermitian form Φ on $V^* = H^1(P_0, L_0)$, unique up to a positive real factor, of signature $(1, |S| - 3)$ such that $\Gamma_\mu \subset PU(V^*, \Phi)$. One has $\Gamma_{\mu,H} \subset PU(V^*, \Phi)$.

COROLLARY 10.17. *Assume that the 5-tuples $_i\mu$ $(i = 1,2,3)$ of 10.8 are ball 5-tuples. Let V_i^* and Φ_i^* denote the above 3-dimensional vector space and hermitian form associated to $_i\mu(i = 1,2,3)$. Then the correspondence between the local systems V_i and V_j sends $PU(V_i^*, \Phi_i)$ to $PU(V_j^*, \Phi_j)$.*

PROOF. $PU(V_i^*, \Phi_i)$ is the set of **R**-points of an algebraic subgroup of $PGL(V_i^*)$ defined over **R**. By [DM] (11.5), $\Gamma_{i\mu}$ is Zariski-dense in $PU(V_i^*)(i = 1,2,3)$. By (10.6), conjugation by an element in $PGL(3\mathbf{C})$ sends $\Gamma_{i\mu,H_i}$ to $\Gamma_{j\mu,H_j}$. Hence it sends $PU(V_i^*, \Phi_i)$ to $PU(V_j^*, \Phi_j)$.

We note that up to a positive real constant, there is a unique hermitian form of signature $(1, 2)$ invariant under $PU(V_i^*, \Phi_i)$. Consequently, the conjugacy of (10.6) sends Φ_1 to a positive multiple of Φ_2. Thus if we identify $PU(V_i^*, \Phi_i)$ with $PU(1, 2)$ in any way, we conclude

COROLLARY 10.18. *Assume that $_1\mu = (a, a, b, b, 2 - 2a - 2b)$, and $_2\mu = (1 - b, 1 - a, a + b - \frac{1}{2}, a + b - \frac{1}{2}, 1 - a - b)$ are ball 5-tuples. Then $\Gamma_{_1\mu,H_1}$ and $\Gamma_{_2\mu,H_2}$ are conjugate in $PU(1, 2)$. If moreover $a = b$ and $_3\mu = (\frac{3}{2} - 2a, a, a, a, \frac{1}{2} - a)$ then $\Gamma_{_1\mu,V}$ is conjugate to $\Gamma_{_3\mu}$ in $PU(1, 2)$ and $\Gamma_{_1\mu,\Sigma_4}$ is conjugate to $\Gamma_{_3\mu,\Sigma_3}$.*

Remark 10.19 In case $_1\mu$ and $_2\mu$ satisfy the integrality condition ΣINT of 8.2, the conjugacy of $_1\Gamma_{\mu,H_1}$ and $_2\Gamma_{\mu,H_2}$ can be deduced more directly by the rigidity theorem of [M1]. For the map φ of (10.3) yields a homeomorphism

$$\varphi : (Q - D_{CF_1} - D_{CF_2})/H_1 \xrightarrow{\sim} Q/H_2.$$

Let $_iQ'$ denote the subset of Q on which H_i operates freely $(i = 1, 2)$. We have

$$\pi_1((Q - D_{CF_1} - D_{CF_2})/H_1, \bar{0}) \simeq \pi_1(_1Q'/H_1, \bar{0})$$

since the two subsets of Q/H_1 differ by a closed subset of **C**-codimension at least two. Similarly

$$\pi_1(Q/H_2, \bar{0}) \simeq \pi_2(_2Q'/H_2, \bar{0}).$$

Consequently

$$\pi_1(_1Q'/H_1, \bar{0}) \simeq \pi_1(_2Q'/H_2, \bar{0}).$$

By Lemma 8.6.1, we see that the kernels of

$$\theta_{H_i} : \pi_1({}_iQ'/H_i, \overline{0}) \rightarrow Aut\ B_{\overline{0}}^{\pm} = PU(1,2)$$

coincide. Consequently $\Gamma_{1\mu,H_1}$ and $\Gamma_{2\mu,H_2}$ are isomorphic lattices in $PU(1,2)$ by [M3]. By the rigidity theorem, they are conjugate.

Remark 10.20 We can get isomorphisms between the linear groups $\Gamma_{1\alpha,H_1}$ and $\Gamma_{2\alpha,H_2}$ for suitable choice of the exponents on divisors at infinity of Q_2''/H_2 which come from finite divisors in Q_1.

Applying 7.16 and 7.22.2, we get:

THEOREM 10.21. *Let $_1\mu, _2\mu$, and ψ be as in 10.5. Let $_2\alpha$ be exponents giving rise to $_2\mu$, i.e., $_2\mu_s = \sum_t \alpha_{s,t}$. Assume that $_2\alpha$ is H_2-invariant and of sum 1. Assume further that*

(10.21.1)
$$(_2\alpha_{FC}, _2\beta_{FC}) = (0, \frac{1}{2}).$$

Then the pullback of $L(_2\alpha)/H_2$ by $\psi = \varphi \circ_1 \pi$ is hypergeometric-like, with exponents given by

(10.21.2)
$$(_1\alpha_{AA}, _1\beta_{AA}) = 2(_2\alpha_{AF}, _2\beta_{AF})$$
$$(_1\alpha_{BB}, _2\beta_{BB}) = 2(_2\alpha_{BF}, _2\beta_{BF})$$
$$(_1\alpha_{AB}, _1\beta_{AB}) = \frac{1}{2}(_1\alpha_{FF}, _2\beta_{FF})$$
$$(_1\alpha_{AC}, _1\beta_{AC}) = (_2\alpha_{AC}, _2\beta_{AC})$$
$$(_1\alpha_{BC}, _1\beta_{BC}) = (_2\alpha_{BC}, \beta_{BC}).$$

PROOF. This theorem is an immediate consequence of (7.22.5).

Note: If $_2\mu$ is H_2-invariant and such that $1 - _2\mu_F - _2\mu_C = \frac{1}{2}$, there exists $_2\alpha$ of sum 1, H_2-invariant, with $_2\alpha_{FC} = 0$, giving rise to it; for the exponents of the Lauricella local system relative to C, A, B and $_2\mu$, satisfy $_2\alpha_{FC} = 0$ and are, after averaging, H_2-invariant.

Remark 10.22 (i) The fact that the system of affine linear equations (10.5.1) and (10.5.2) has a solution $_1\mu$ is not a priori clear. In a similar vein, it follows from 10.21 that if $_2\mu$ is H_2-invariant, of sum 1, and satisfies

(10.21.1), then there exist $(_1\alpha, {}_1\beta)$ defined by (10.21.2) such that $_1\alpha$ is of sum 1, H_1-invariant, and that

$$_1\beta_{s,t} = \sum_{\{s,t\}\cap\{u,v\}=\phi} {}_1\alpha_{u,v}.$$

In a first version of this paper, we checked this directly; it appeared miraculously after an elementary computation and was explained ultimately by Theorem 7.11.

(ii) The space of H_1-invariant exponents $_1\alpha$ with sum 1, as well as the space of H_2-invariant exponents $_2\alpha$ with sum 1 satisfying (10.21.1), are affine spaces of the same dimension (4). By 7.17 (ii), the equations (10.21.2) establish an isomorphism between those affine spaces.

It follows that any H_1-invariant hypergeometric-like local system L of holomorphic functions on Q corresponds by φ to an H_2-invariant hypergeometric-like local system with $\alpha_{FC} = 0, \beta_{FC} = \frac{1}{2}$. One could try to check this directly, by applying (7.1) to $(\varphi_2^{-1}\tau)^*(L/H_1)$. A difficulty is that the divisor D_{AB} of Q^+ has been contracted to a point in the model $\overline{Q}_2$.

(iii) Let $_i\mu$ be deduced from $_i\alpha$ by the usual rule

$$_i\mu_u = \sum_v {}_i\alpha_{u,v}.$$

Then, for $_1\alpha$ corresponding to $_2\alpha$ as in 10/17. $_1\mu$ depends only on $_2\mu$. Indeed, one has

$$1 -_i \mu_s -_i \mu_t =_i \beta_{s,t} -_i \alpha_{s,t}$$

and by 10.21.2, each $1 -_1 \mu_s -_1 \mu_t$ is a fixed multiple of some $1 -_2 \mu_s -_2 \mu_t$. One could also argue that μ detects the hypergeometric-like local system $L(\alpha)$ modulo twisting, and that invertible H_1-invariant functions on Q correspond by φ to invertible H_2-invariant functions on Q with valuation 0 along D_{FC}.

§11. ANOTHER ISOGENY

11.1. Fix $S = \{1,2,3,4,5\}$ and let $\Sigma(3)$ be the subgroup of $\Sigma(S)$ fixing 4 and 5. With the notations of 10.1 for $N = 5$, we will consider the compactifications of Q corresponding to the following T:

$_1\bar{Q} : T = \phi$

$_2\bar{Q} : T$ reduced to $\{4,5\}$

$\bar{Q} : T = \{\{1,5\}, \{2,5\}, \{3,5\}, \{4,5\}\}$.

The first part of this section (up to 11.19), is devoted to construction of a map

$$\varphi : {}_1\bar{Q}/\Sigma(3) \longrightarrow {}_2\bar{Q}/\Sigma(3)$$

to which we will apply 7.18 to obtain an injection $\Gamma_{1\mu,\Sigma(3)} \to \Gamma_{2\mu,\Sigma(3)}$ where, for $\frac{1}{\rho} + \frac{1}{\sigma} = \frac{1}{6}$,

$$_1\mu = \mu(\rho, 3, \sigma) = \left(\frac{1}{2} - \frac{1}{\rho}, \frac{1}{2} - \frac{1}{\rho}, \frac{1}{2} - \frac{1}{\rho}, \frac{1}{6} + \frac{1}{\rho}, 2\left(\frac{1}{6} + \frac{1}{\rho} \right) \right)$$

$$_2\mu = \mu(3, \rho, \sigma) = \left(\frac{1}{6}, \frac{1}{6}, \frac{1}{6}, \frac{5}{6} - \frac{1}{\rho}, \frac{5}{6} - \frac{1}{\sigma} \right)$$

The map φ will be deduced by blowing up from a map, also denoted by φ, from $\bar{Q}/\Sigma(3)$ to $\bar{Q}/\Sigma(3)$:

$$
\begin{array}{ccc}
{}_1\bar{Q}/\Sigma(3) & \longrightarrow & {}_2\bar{Q}/\Sigma(3) \\
\downarrow & & \downarrow \\
\bar{Q}/\Sigma(3) & \longrightarrow & \bar{Q}/\Sigma(3)
\end{array}
$$

In contrast to the preceding cases, we do not have a modular description of φ; we define it in terms of explicit coordinates on $Q/\Sigma(3)$.

The space Q is the quotient by $PGL(2)$ of the space of 5-tuples of distinct points (A_1, A_2, A_3, B, C) on a projective line P. The quotient $Q/\Sigma(3)$ can be similarly defined, with A_1, A_2, A_3 unordered. The compactification $\bar{Q}$ of

Q is the one where one allows any of A_1, A_2, A_3 and B to coalence - but not all four at the same point, and where C is not allowed to coincide with any of A_1, A_2, A_3 or B; it is Q^{st} for weights $\left(\frac{1}{4} + \varepsilon, \frac{1}{4} + \varepsilon, \frac{1}{4} + \varepsilon, \frac{1}{4} + \varepsilon, 1 - 4\varepsilon\right)$ with $0 < \varepsilon < \frac{1}{12}$. It is isomorphic to $\mathbf{P}^2(\mathbf{C})$ by the following map: given (A_1, A_2, A_3, B, C), choose a coordinate z with $z(B) = 0, z(C) = \infty$ and take the point of $\mathbf{P}^2(\mathbf{C})$ with homogeneous coordinates $(z(A_1), z(A_2), z(A_3))$.

Set $R = \bar{Q}/\Sigma(3)$. Let D_{AA} (resp. D_{AB}) be the image in R of the divisor $D_{A_iA_j}$ (resp. D_{A_iB}) on $\bar{Q}$. The construction of φ can be summarized as follows

PROPOSITION 11.2.. (i) *The space R, taken together with the two divisors D_{AA} and D_{AB}, has no nontrivial automorphism. R has two singular points; let R' be their complement.*

(ii) *The biggest connected covering of $R' - D_{AA}$, with ramification indices along D_{AA} dividing 3, is an $A(4)$-covering. Let R_{12} be its completion above R.*

(iii) *The space R_{12} is a projective quadratic cone. The inverse image of D_{AA} consists of four generatrices with multiplicity three, permuted by $A(4)$. Denote one of them D'_{AA} and let C_3 denote its stabilizer in $A(4)$; C_3 fixes D'_{AA} pointwise. The inverse image D'_{AB} of D_{AB} is irreducible.*

(iv) *The quotient $R_4 := R_{12}/C_3$, taken together with D'_{AB}/C_3 and D'_{AA}/C_3 on it, is isomorphic (uniquely by (i)) to (R, D_{AA}, D_{AB}):*

$$(R, D_{AA}, D_{AB}) \sim (R_4, D'_{AB}/C_3, D'_{AA}/C_3).$$

The map φ is then defined as

$$R \underset{11.2(iv)}{\longrightarrow} R_4 \to R.$$

The proof of 11.2 will occupy us until 11.15. The figures in 11.16 may help one follow the arguments.

11.3. The space R is the quotient by $PGL(2)$ of the space of 5-tuples (A_1, A_2, A_3, B, C), A_i's unordered, with coalescence allowed as in 11.1. We fix $B = 0, C = \infty$; R then becomes the quotient by $\mathbf{C}^*$ of the space of unordered triples of numbers $a_1, a_2, a_2 \in \mathbf{C}$, with $(a_1, a_2, a_3) \neq (0, 0, 0)$. Let σ_i be the elementary symmetric functions in the a's. The map $(a_1, a_2, a_3) \mapsto (\sigma_1, \sigma_2, \sigma_3)$ identifies R with the space of triples $(\sigma_1, \sigma_2, \sigma_3) \neq (0, 0, 0)$, divided by $\mathbf{C}^*$ acting by $(\sigma_1, \sigma_2, \sigma_3) \mapsto (\lambda\alpha_1, \lambda^2\alpha_2, \lambda^3\sigma_3)$. This is the weighted homogeneous projective plane $\mathbf{P}(1, 2, 3)$

$$\mathbf{P}(1, 2, 3) = \text{Proj } \mathbf{C}[\sigma_1, \sigma_2, \sigma_3]$$

with σ_i of weight i.

The equation of the divisor D_{AA} is the discriminant of the equation
$$T^3 - \sigma_1 T^2 + \sigma_2 T - \sigma_3 = 0$$

$$D_{AA} : \text{Discr}(-\sigma_1, \sigma_2, -\sigma_3) = 0.$$

The equation of D_{AB} is
$$D_{AB} : \sigma_3 = 0.$$

11.4. Let us recall some facts about weighted projective spaces, and $\mathbf{P}(1, 2, 3)$ in particular.

(11.4.1) The weighted projective space $\mathbf{P}(d_0, \ldots, d_n)$ $(d_i \geq 1)$ is the quotient of
$\mathbf{C}^{n+1} - \{0\}$ by $\mathbf{C}^*$ acting by $(x_0, \ldots, x_n) \mapsto (\lambda^{d_0} x_0, \ldots, \lambda^{d_n} x_n)$. It is the Proj of the graded polynomial algebra with generators x_i of weight d_i. At the points where the non vanishing x_i have relatively prime weights, it is non singular and $\mathcal{O}(1)$ is a line bundle. At the points where the gcd of the weights of the non vanishing x_i divides d, $\mathcal{O}(d)$ is a line bundle and $\mathcal{O}(dk) = \mathcal{O}(d)^{\otimes k}$.

In general, it is best to treat weighted homogeneous projective spaces as orbifolds. However, in our case of $\mathbf{P}(1, 2, 3)$, we can do as well by treating it as a normal surface.

(11.4.2) The surface $\mathbf{P}(1, 2, 3) = \text{Proj } \mathbf{C}[\sigma_1, \sigma_2, \sigma_3]$ has two singular points: $(0, 1, 0)$ and $(0, 0, 1)$, of type A_1 and A_2 respectively. At $(0, 1, 0)$ the singularity is that of $\mathbf{C}^2$ divided by the involution $(x, y) \mapsto (-x, -y)$: a quadratic cone; at $(0, 0, 1)$, the singularity is that of $\mathbf{C}^2$ divided by μ_3 acting by $(x, y) \mapsto (\zeta x, \zeta^{-1} y)$. This description also tells how $\mathbf{P}(1, 2, 3)$ should be treated as an orbifold.

(11.4.3) Let $\mathbf{P}(1, 2, 3)' := \mathbf{P}(1, 2, 3)$ - singular points. Then,

$$\text{Pic}(\mathbf{P}(1, 2, 3)') = \mathbf{Z}$$

and is generated by $\mathcal{O}(1)$. The line bundle $\mathcal{O}(d)$ extends at the singular point $(0, 1, 0)$ (resp. $(0, 0, 1)$) if and only if $2 \mid d$ (resp. $3 \mid d$). The projective coordinate ring is

$$\oplus H^0(\mathbf{P}(1, 2, 3)', \mathcal{O}(n)).$$

(11.4.4) The *weight* of a Weil divisor D is the class in Pic of its trace on $\mathbf{P}(1, 2, 3)'$. For two divisors D_1, D_2, one has

$$D_1 \cdot D_2 = \frac{1}{6} w(D_1) w(D_2).$$

The intersection number is taken in the sense of intersections on normal surfaces, or orbifolds.

(11.4.5) There is on $\mathbf{P}(1, 2, 3)$ exactly one divisor of weight one: $\sigma_1 = 0$. It contains the two singular points. The complement of this unique divisor is an affine plane, with coordinates $X = x/\sigma_1^2$, $Y = y/\sigma_1^3$ where x (resp. y) is a weight two (resp. three) generator of the projective coordinate ring. The coordinate X (resp. Y) is unique modulo $X \mapsto \lambda X + \mu$ with $\lambda \neq 0$ (resp. $Y \mapsto \alpha Y + \beta X + \gamma$ with $\alpha \neq 0$).

In these coordinates, the weight of the divisor $P(X, Y) = 0$ is the weight of P, with $w(X) = 2$ and $w(Y) = 3$. In particular, the straight lines are of weight 2, if of the form $X = $ constant, and of weight 3 otherwise.

PROPOSITION 11.5. *(i): On R (identified in 8.3 with* $\mathbf{P}(1, 2, 3)$*), D_{AA} and D_{AB} are irreducible of weight 6 and 3 respectively. The divisors D_{AA} and D_{AB} meet at two points: a transverse intersection (image of $A_1 = A_2, A_3 = B$ in $\bar{Q}$) and a simple tangency point (image of $A_1 = A_2 = B$ in $\bar{Q}$).*

(ii) There is a unique system of affine coordinates X, Y as in (11.4.5) for which the equations of D_{AA} and D_{AB} are:

$$D_{AA} : Y^2 - X^3 = 0$$

$$D_{AB} : (Y - 1) = \frac{3}{2}(X - 1)$$

(tangency at $(1, 1)$, transverse intersection at $(1/4, -1/8)$).

PROOF. Let us rewrite the polynomial

$$T^3 - \sigma_1 T^2 + \sigma_2 T - \sigma_3$$

as

$$(T - \sigma_1/3)^3 - p(T - \sigma_1/3) + q$$

with p and q isobaric of weight 2 and 3 respectively in $\sigma_1, \sigma_2, \sigma_3$. The discriminant equation (D_{AA}) is then the weight 6 equation

$$\left(\frac{p}{3}\right)^3 - \left(\frac{q}{2}\right)^2 = 0,$$

while the equation of D_{AB} (11.3) is of weight 3.

We claim that the divisors D_{AA} and D_{AB} intersect exactly as described in (i). For in $\bar{Q}$, the divisors $A_1 = A_2$ and $A_3 = B$ meet transversely. Locally, to go to R, one has to divide by an involution fixing $A_1 = A_2$ and stabilizing $A_3 = B$. The intersection stays transversal in R. The other intersection point is hence a simple tangency for the total intersection number $D_{AA} \cdot D_{AB}$ is $3 = 1 + 2$ by (11.4.4). If the tangency point is at $p/\sigma_1^2 = p_0, q/\sigma_1^3 = q_0$, coordinates as in (ii) are provided by $X = p/p_0\sigma_1^2, Y = q/q_0\sigma_1^3$.

The uniqueness assertion in 11.5 (ii), is given by:

PROPOSITION 11.6.. *Let (M, D, E) be a system as follows*

(i) *M is isomorphic to $\mathbf{P}(1, 2, 3)$; let F denote its unique divisor of weight 1.*

(ii) *D is an irreducible divisor of weight 6 on M whose trace on the affine plane $M - F$ is a cuspidal cubic.*

(iii) *E is an irreducible divisor of weight 3 on M whose trace on the affine plane $M - F$ is a straight line tangent to D at a non singular point.*

Then, there is a unique isomorphism of (M, D, E) with (R, D_{AA}, D_{AB}).

PROOF. We may assume that $M = \mathbf{P}(1, 2, 3)$. We have to show that there is a unique system of affine coordinates (X, Y) as in (11.4.5) such that the equations of D and E are as in 11.5 (ii).

We first normalize (X, Y) to have the cusp at $(0, 0)$. The tangent at the cusp is not $X = 0$, because the intersection number of D (weight 6) and $X = 0$ (weight 2) is only 2, not 3. We can hence normalize (X, Y) further and get that tangent to be $Y = 0$. The equation is then $aX^3 + bY^2 = 0$ (lower weight terms have been excluded). The final normalization, that makes the tangency point $(1, 1)$, determines X and Y and leads to the equations in 11.5 (ii).

11.7. Remark (i) in our case, for p and q as in the proof of 11.5, X and Y are

$$X = \left(\frac{p}{3}\right) \Big/ \left(\frac{-\sigma_1}{3}\right)^2$$

$$Y = \left(\frac{q}{2}\right) \Big/ \left(\frac{-\sigma_1}{3}\right)^3.$$

We define $x = \sigma_1^2 X, y = \sigma_1^3 Y$.

(ii) With x and y as above, the singular points are given by $(\sigma_1, x, y) = (0, 1, 0)$ and $(0, 0, 1)$; none is contained in D_{AA} while the first (type A_1) is on D_{AB}, corresponding to triples $(a_1, a_2, a_3) = (a, 0, -a)$ on $\bar{Q}$.

11.8. Fix X, Y as in 11.5 (ii) and consider the ramified covering of the affine (X, Y)-plane obtained by extracting the cube root of $Y^2 - X^3$. Outside of the singular point of type A_2, its completion R_3 above R can be described as follows. Define $x = \sigma_1^2 X, y = \sigma_1^3 Y$. The equation $y^2 - x^3$ of D_{AA} is a section of the line bundle $\mathcal{O}(6)$. At a general point P of D_{AA}, if one chooses a local trivialization of $\mathcal{O}(6)$, $y^2 - x^3$ becomes a function with non vanishing differential at P: it vanishes simply on D_{AA}. The line bundle $\mathcal{O}(6)$ is $\mathcal{O}(2)^{\otimes 3}$, $y^2 - x^3$ is a global section of $\mathcal{O}(6)$, and R_3, contained in the total space of $\mathcal{O}(2)$, is the space of cube roots of $y^2 - x^3$. On R_3, the pullback of $\mathcal{O}(2)$ has a section d whose cube is the inverse image of $y^2 - x^3$, and (R_3, d) is universal for this property. As $y^2 - x^3$ vanishes simply along D_{AA}, outside of the singular point of type A_2 the triple covering R_3 of R ramifies exactly along D_{AA}, with ramification index 3.

The space R_3 can also be viewed as

$$\text{Proj } \mathbf{C}[\sigma_1, x, y, d]/(d^3 + x^3 - y^2),$$

with generators of degree $1, 2, 3$ and 2. Locally on this Proj, one of σ_1, x, y, does not vanish, so that the map to R is defined.

The map $R_3 \to R$ in fact ramifies at the A_2-singular point, above which R_3 is non singular; we do not use this.

The equation $d^3 = y^2 - x^3$ defining the projective coordinate ring of R_3 can be rewritten

(11.8.1) $$y^2 = (x+d)(x+\zeta d)(x+\zeta^2 d)$$

for ζ a primitive cube root of 1, and $(x+d), (x+\zeta d), (x+\zeta^2 d)$, of weight two, are related by the single linear relation

$$(x+d) + \zeta(x+\zeta d) + \zeta^2(x+\zeta^2 d) = 0.$$

Let us now extract the square roots u, v, w of $x+d$, $x+\zeta d$, $x+\zeta^2 d$, with $uvw = y$:

$$R_{12} := \text{Proj } \mathbf{C}[\sigma_1, x, y, d, u, v, w]/I$$

with generators of weight $1, 2, 3, 2, 1, 1, 1$ and I is generated by the relations

$$y = uvw$$
$$u^2 = x+d, v^2 = x+\zeta d, w^2 = x+\zeta^2 d.$$

One has $3x = u^2 + v^2 + w^2$, $3d = u^2 + \zeta^{-1}v^2 + \zeta^{-2}w^2$ and the projective coordinate ring is more simply

(11.8.2) $$\mathbf{C}[\sigma_1, u, v, w]/(u^2 + \zeta v^2 + \zeta^2 w^2)$$

with generators of weight 1. Thus R_{12} is a projective quadratic cone, with the pullback of $\mathcal{O}(1)$ from R to R_{12} giving rise to the embedding in $\mathbf{P}^3$.

By (11.8.1), R_{12} is an etale covering of R_3, except above the points $x = d = y = 0$ (where the divisors $x+d = 0, x+\zeta d = 0, x+\zeta^2 d = 0$ meet) and above the three points $\sigma_1 = y = 0$, where $\mathcal{O}(1)$ is not a line bundle. The latter three points are the inverse image of the A_1-singularity of R and above them R_{12} is non singular.

The ramified covering R_{12} of R is Galois; its Galois group consists of cyclic permutations of u, v, w with an even number of sign changes. A cyclic permutation of u, v, w maps d to ζd with $\zeta \in \mu_3$. The Galois group is $A(4)$. There is ramification only at the singular points of R, and along D_{AA}, where the ramification index is 3. The quadratic cone R_{12} is simply connected, and hence R_{12} is the universal covering of R with the prescribed ramification. This proves 11.2 (ii).

11.9. The equation for the inverse image of D_{AA} in R_{12} is $d^3 = 0$. One has $3d = u^2 + \zeta^{-1}v^2 + \zeta^{-2}w^2$, and $d = 0$ in R_{12} is the intersection of the two quadratic cones in $\mathbf{P}^3$ with the same vertex $(\sigma, u, v, w) = (1, 0, 0, 0)$

$$u^2 + \zeta v^2 + \zeta^2 w^2 = 0 \qquad (R_{12})$$
$$u^2 + \zeta^{-1}v^2 + \zeta^{-2}w^2 = 0 \qquad (d = 0).$$

These two cones intersect in four generatrices, corresponding to the four points $(1, 1, 1)(-1, 1, 1)(1, -1, 1)(1, 1, -1)$ of the projective conic $u^2 + \zeta v^2 + \zeta^2 w^2 = 0$. Let D'_{AA} be the one corresponding to $(1, 1, 1)$. It is fixed by the cyclic subgroup C_3 of order 3 of $A(4)$ permuting u, v, and w.

The inverse image of D_{AB} in R_3 is irreducible: it ramifies above D_{AB} at the two points of $D_{AA} \cap D_{AB}$. Let us show that the inverse image D'_{AB} of D_{AB} in R_{12} is also irreducible. If not, its components, permuted by $A(4)$, and permuted transitively by $\mathrm{Gal}(R_{12}/R_3) = \mathbf{Z}/2 \times \mathbf{Z}/2$, would number four since the stabilizer of a component is a subgroup of $A(4)$ of index dividing 4. They form the intersection of the quadratic cone R_{12} with the cubic surface $\sigma_3 = 0$. This is a projective curve of degree 6, and $4 \nmid 6$. This proves 11.2 (iii).

11.10 Remark.   The four points $(1, 1, 1)(-1, 1, 1), (1, -1, 1)$ $(1, 1, -1)$ on the projective conic $u^2 + \zeta v^2 + \zeta^2 w^2$ are stable by an action of $A(4)$, hence form an equiharmonic quaternary (the cross ratio is a primitive sixth root of 1).

LEMMA 11.11. *the quotient $R_4 := R_{12}/C_3$ is isomorphic to $\mathbf{P}(1, 2, 3)$. The inverse image of $\mathcal{O}(2)$ is $\mathcal{O}(1)$ on the quadratic cone R_{12}.*

PROOF. Let us extract the cube root of the projective coordinate σ_3 of $\mathbf{P}(1, 2, 3) = \mathrm{Proj}\ \mathbf{C}[\sigma_1, \sigma_2, \sigma_3]$. We get

$$\tilde{R}_3 := \mathrm{Proj}\ \mathbf{C}[\sigma_1, \sigma_2, a]$$

with generators of weight $1, 2, 1$ ($\sigma_3 = a^3$). The covering $\tilde{R}_3$ is ramified only at the singular points and along the divisor $\sigma_3 = 0$, along which the ramification index is 3. The surface $\tilde{R}_3$ is also the Proj of the even weight part of $\mathbf{C}[\sigma_1, \sigma_2, a]$: generators $\sigma_1^2, a^2, \sigma_1 a, \sigma_2$, only relation $\sigma_1^2 \cdot a^2 = (\sigma_1 a)^2$.

It is a quadratic cone with its $\mathcal{O}(1)$ the inverse image of $\mathcal{O}(2)$. The inverse image of $\sigma_3 = 0$ is the generatrix $a^2 = \sigma_1 a = 0$.

To conclude the proof, we need the following lemma.

LEMMA 11.12. *Any two cyclic* $\mathbf{Z}/3$ *groups of automorphisms of a quadratic cone fixing a generatrix are conjugate.*

PROOF. All generatrices being conjugate under the automorphisms of the cone, we may assume that the same one, g, is fixed by both groups. Let A be the group of automorphisms fixing g, and $\bar{A}$ the quotient of A which acts faithfully on the set of generatrices (a conic). The group $\bar{A}$ is the affine group (fixing one point on a curve of genus 0): $\bar{A} = \mathbf{G}_m \times \mathbf{G}_a$ and is the quotient of A by a unipotent group.

Two finite subgroups in the linear group A are conjugate if and only if their image in A modulo its unipotent radical are. We have $A/R_u A \simeq \mathbf{G}_m$. Since $\mathbf{G}_m$ contains only one cyclic subgroup of a given order, 11.12 follows.

We now finish the proof of 11.11. We have already seen that $\mathbf{P}(1,2,3)$ is the quotient of R_{12} by some order 3 group of automorphisms fixing a generatrix. By the lemma, it is conjugate to C_3 and 11.11 follows.

11.13. On the quadratic cone R_{12}, the divisor D'_{AA} is of weight 1/2: twice a generatrix of a quadratic cone is the intersection of the cone with a tangent hyperplane; hence $\mathcal{O}(2D'_{AA}) \sim \mathcal{O}(1)$. The divisor D'_{AB} is of weight 3. Dividing by C_3, and taking the ramification into account, we get that the irreducible divisors D'_{AA}/C_3 and D'_{AB}/C_3 on $R_4 = R_{12}/C_3 = \mathbf{P}(1,2,3)$ are of weight 3 and 6 respectively.

The divisor D'_{AB} meets each of the four generatrices above D_{AA} in two points. Above the transverse intersection point of D_{AA} and D_{AB}, the intersection is transversal. Above the tangency point of D_{AA} and D_{AB}, D'_{AB} has a cusp.

Dividing by C_3, we get that D'_{AB}/C_3 has a cusp (image of the three cusps of D'_{AB} not on D'_{AA}) and that D'_{AB}/C_3 and D'_{AA}/C_3 meet in two points (a transversal intersection image of the transversal intersection, and a tangency). Applying 11.6, we get 11.2 (iv).

11.14 Remark. The proof of 11.11 and 11.13 shows that $R_{12} \to R_4 \simeq R$ is etale except above the singular points of R and D_{AB}, and this covering has ramification index 3 along D_{AB}. As the smooth locus of R_{12} is simply connected, R_{12} is the largest covering with those ramification properties.

11.15 Remark. The ramification of

$$\varphi : R \xrightarrow{\ \xi^{-1}\ } R_4 \to R$$

where ξ denotes the inverse of the isomorphism $R \to R_4$ of 11.2 (iv) is as follows.

There is ramification above the two singular points, and along one of the two irreducible components of the inverse image of D_{AA} with ramification index 3. The other component over D_{AA} is D_{AB}. There is no other ramification. In particular, φ is etale at $D_{AA} \cap D_{AB}$.

11.16. R is the quotient of $\bar{Q}(\simeq \mathbf{P}^2)$ by $\Sigma(3)$:

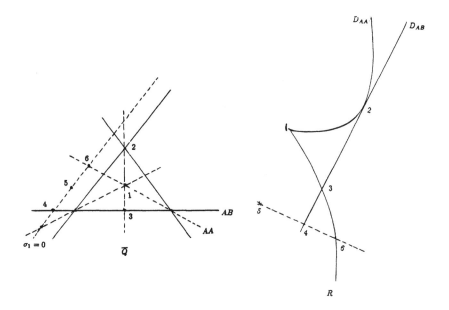

On $\bar{Q}$: solid lines: $A_i = B$; dashed lines: $A_i = A_j$; dotted line: $\sigma_1 = 0$. In a correct figure, 3 and 4 would be harmonic conjugates, and there would be two points 5, complex conjugates of each other. On R: the numbered points are the images of the points on $\bar{Q}$ with the same label. The dotted line is $\sigma_1 = 0$, 4 is the A_1 singularity and 5 the A_2 singularity.

The quadratic cone R_{12} maps to $R \simeq R_{12}/A(4)$ by ψ_{12} and to $R \simeq R_4 = R_{12}/C_3$ by $\xi\psi_3$. The inverse image $\psi_{12}^{-1}(D_{AA})$ breaks into four generatrices, one of which, D'_{AA}, is fixed by C_3 and maps to D_{AB} in R via $\xi\psi_3$. The inverse image $D'_{AB} = \psi_{12}^{-1}(D_{AB})$ has four cusps. Its image in R via $\xi\psi_3$ is D_{AA}, with three of the cusps of D'_{AB} mapping to the cusp of D_{AA}.

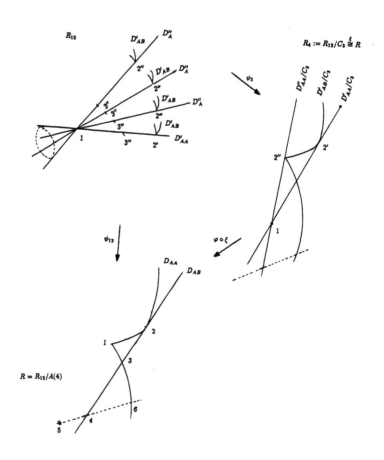

On R_{12}/C_3, the A_1 singularity is the point 1, image of the vertex of the cone. The new weight one curve is the image of the only generatrix other than D'_{AA} stabilized by C_3. The A_2-singular point on it is the image of the point of that generatrix fixed by C_3, other than the vertex of the cone (intersection of that generatrix with the old $\sigma_1 = 0$: the only smooth plane section stabilized by C_3).

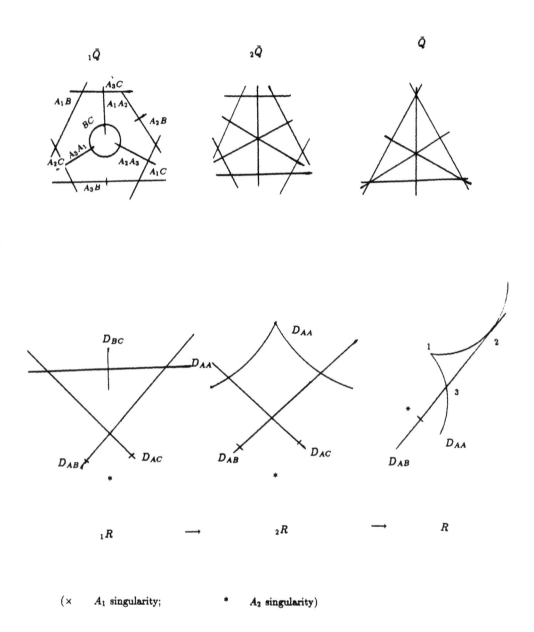

($\times$ A_1 singularity;   * A_2 singularity)

11.17. With the notations of 11.1, $_1\bar{Q}$ can be obtained from the projective plane $\bar{Q}$ by blowing up the points $A_i = A_j = B$ $(i \neq j)$ and $A_1 = A_2 = A_3$, while $_2\bar{Q}$ is obtained by blowing up only the first three of those points.

Define

$$_iR =_i \bar{Q}/\Sigma(3) \qquad (i = 1, 2).$$

In 11.18 (resp. 11.19), we will describe $_1R$ (resp. $_2R$) as the blow up of R at an ideal I. Let us recall that such blow up is Proj of the sheaf of graded rings $\bigoplus\limits_{n \geq 0} I^n$.

We will repeatedly use the following. Let $f : X \to Y$ be a morphism, and let $Z \subset Y$ be a closed subscheme defined by an ideal I. Let $\tilde{Y}$ be the blow up of Y along Z (i.e. at I). Then, if the scheme $f^{-1}(Z)$ is a Cartier divisor in X (i.e., given by a sheaf of invertible ideals), then f lifts (uniquely) to a morphism into $\tilde{Y}$.

11.18. The compactification $_2\bar{Q}$ is deduced from $\bar{Q}$ by blowing up the three points $A_i = A_j = B$. Dividing by $\Sigma(3)$, we get $_2R$. Let P be the tangential intersection point of D_{AA} and D_{AB} in R; its inverse image in $_2R$ is an irreducible curve C and $_2R - C \to R - P$ is an isomorphism. More precisely, we claim that to get $_2R$ from R, one has to blow up the ideal defining, at that point, the scheme-theoretic intersection of D_{AA} and D_{AB}. Let us check this in local coordinates. Near the point $P' : A_1 = A_2 = B$, one can find on $\bar{Q}$ local coordinates (x, y) centered at P' such that the lines $A_1 = A_2, A_i = B$ and the involution $\sigma : A_1 \leftrightarrow A_2$ are given by

$$A_1 = A_2 : y = 0$$
$$A_i = B : y = \pm x$$
$$A_1 \leftrightarrow A_2 : (x, y) \mapsto (x, -y).$$

Dividing by σ, we get a space etale above $P \in R$, and (x, y^2) descend to local coordinates (X, Y) on R at P. In those coordinates, the tangent curves D_{AA} and D_{AB} are

$$D_{AA} : Y = 0$$
$$D_{AB} : Y = X^2$$

and the ideal defining their intersection is (X^2, Y). The inverse image of this ideal is (x^2, y^2). It becomes invertible if one blows up (x, y) and the map

$$(x, y) \mapsto (x, y^2) : \mathbf{C}^2 \to \mathbf{C}^2$$

extends to a map

$\mathbf{C}^2$ with (x, y) blown up $\to \mathbf{C}^2$ with (X^2, Y) blown up.

The map we obtain does not blow down any curve, and since σ stabilizes the blown up ideal, the above map factors through a finite birational map

$\mathbf{C}^2$ with (x, y) blown up$/\sigma \to \mathbf{C}^2$ with (X^2, Y) blown up

which is an isomorphism by Zariski's main theorem.

By the definition of $_2R$, B and C play symmetric roles, leading to the above picture for the map $_2R \to R$.

11.19. The compactification $_1\bar{Q}$ is obtained by the additional blowing up of the point $A_1 = A_2 = A_3$. Dividing by $\Sigma(3)$, we get $_1R$ a blow up of $_2R$, with center at the cusp P of D_{AA}. More precisely, we claim that to get $_1R$ from $_2R$, one has to blow up the ideal of functions which, at that point, on the normalization of the cusp of D_{AA}, vanish of order ≥ 6. Let us check this in local coordinates. Near the point $P' : A_1 = A_2 = A_3$, one can find on $\bar{Q}$ local coordinates (x_1, x_2, x_3) with the constraint $x_1 + x_2 + x_3 = 0$, such that the lines $A_i = A_j$ are given by the equations $x_i = x_j$, and the action of $\Sigma(3)$ by permutation of the x's. The elementary symmetric functions σ_2, σ_3 descend as local coordinates on R. Put $x := \sigma_2/3$, $y := \sigma_3/2$. The equation of D_{AA} is then $y^2 = x^3$, and the announced ideal to blow up is (x^3, y^2). Its inverse image becomes invertible after the blowing up of the point $x_1 = x_2 = x_3 = 0$ and one concludes the proof as in 11.18.

PROPOSITION 11.20. *Viewed as a rational map, $\varphi : R \to R$ extends as a finite map, still denoted by φ, from $_1R$ to $_2R$.*

PROOF. We have to check that the two blow ups made of R match, at the two points above the tangential intersection of D_{AA} and D_{AB}. They are (cf. diagram in 11.16):

(a) the point (2) of tangential intersection of D_{AB} and D_{AA}; its blow up is denoted D_{AC}. There $\varphi : R \to R$ is etale and the claim is clear.

(b) The cusp of D_{AA}; its blow up is denoted D_{BC}. Above that cusp, $R_{12} \to R = R_{12}/C_3$ is etale and one can replace φ by the map $R_{12} \to$

R around a cusp point $(2'')$ of $D'_{AB} \subset R_{12}$ not on D'_{AA}. On R, let us choose local coordinates (x, y) around the point (2) as in 11.18 such that the equations of D_{AA} and D_{AB} are $y = 0$ and $y = x^2$ respectively. Then $y^{1/3}$ and x are local coordinates around the point $(2'')$ on R_{12}. Put $(X, Y) = (x, y^{1/3})$. The equation of D'_{AB}, the inverse image in R_{12} of D_{AB}, is $X^2 = Y^2$. The inverse image of the ideal (x^2, y) is (X^2, Y^3). The map extends to the corresponding blow up, not contracting any curve, and as before, this checks 11.20. On the exceptional curves, (generically: coordinates X^2/Y^3 and x^2/y), the map is birational; and with respect to coordinates $(T, U) := (Y^3/X^2, X/Y)$ around a generic point of the exceptional curve in R_{12} and coordinates $(y/x^2, x^3/y)$ in R, the map is $(T, U) \to (T, U^3)$; i.e., φ has ramification index three along the exceptional curve. Above the component of $\varphi^{-1}(D_{AA})$ other than D'_{AA}/C_3 the ramification index of φ is also three. This proves:

COROLLARY 11.21. *The map* $\varphi : {}_1R \to {}_2R$ *ramifies only at isolated points, along* D_{BC}, *and along the "finite" (7.15 (b)) divisor* $D := \varphi^{-1}(D_{AA}) \cap (Q/\Sigma(2))$ *(= the image of* D''_{AA} *in* R_{12}; *cf. second diagram of 11.16), where the ramification index is 3. Divisors are mapped as follows:*

finite divisor D $A = B$ $A = A$ $A = C$ $B = C$

$\downarrow$ $\downarrow$ $\downarrow$ $\downarrow$ $\downarrow$

$A = A$ $A = A$ $A = B$ $A = C$ $A = C$

THEOREM 11.22. *Let*

$$_1\mu = \mu(\rho, 3, \sigma) := \left(\frac{1}{2} - \frac{1}{\rho}, \frac{1}{2} - \frac{1}{\rho}, \frac{1}{2} - \frac{1}{\rho}, \frac{1}{6} + \frac{1}{\rho}, 2\left(\frac{1}{6} + \frac{1}{\rho} \right) \right)$$

$$_2\mu = \mu(3, \rho, \sigma) := \left(\frac{1}{6}, \frac{1}{6}, \frac{1}{6}, \frac{5}{6} - \frac{1}{\rho}, \frac{5}{6} - \frac{1}{\sigma} \right),$$

where $\rho^{-1} + \sigma^{-1} = 6^{-1}$. *Then modulo conjugacy in* $PGL(3, \mathbb{C})$,

(i) $\Gamma_{1\mu, \Sigma(3)} \to \Gamma_{2\mu, \Sigma(3)}$ *with image of index dividing 4.*

(ii) *The image is onto if* σ *is an integer not divisible by 3.*

Proof of (i) We will apply 7.19, completed by 7.20 (i), and 7.21 to the map $\varphi :_1 R \to_2 R$ of 11.21. In 7.15, one considers the composite

$$_1\bar{Q} \to_2 \bar{Q}/\Sigma(3),$$

and $_1Q''$ and $_2Q''$ are obtained from $_1\bar{Q}$ and $_2\bar{Q}$ by deleting finitely many points. Deleting enough of them, one gets rid of any exceptional point for φ. The conditions 7.15 (a)(b)(c) hold [for (c) (ii), because $\frac{2}{3} \notin \mathbf{Z}$]. By 7.18 (d')(e'), we have to consider $\Sigma(3)$-invariant 5-uples $_2\mu$ with sum 2, denoted $(\mu_A, \mu_A, \mu_A, \mu_B, \mu_C)$, with $1 - \mu_A - \mu_A = \frac{2}{3}$. For a general such 5-uple, the non integrality condition 7.18 (f')=7.15 (f) is satisfied. One can apply 7.22.1. We can use 7.22.1 (b) for D_{AA}, as image of D_{AB}; for the other divisors at infinity of $_2R$, we use 7.22.1 (a): on the space of triples (μ_A, μ_B, μ_C), none of the linear forms $\mu_A + \mu_B, \mu_A + \mu_C$ is a linear combination of the forms $3\mu_A + \mu_B + \mu_C$ and $\mu_A + \mu_A$. By 7.20 (i), one may omit the assumption 7.18 (f') in 7.16, and by 7.20 (iii) for any $_2\mu$ as above there is a $\Sigma(3)$-invariant $_1\mu$ with sum 2 related to it by

$$1 -_1\mu_A -_1\mu_B = \frac{1}{2}(1 -_2\mu_A -_2\mu_A) = \frac{1}{3}$$
$$1 -_1\mu_A -_1\mu_A = 2(1 -_2\mu_A -_2\mu_B)$$
$$1 -_1\mu_A -_1\mu_C = 1 - \mu_A - \mu_C$$
$$1 -_1\mu_B -_1\mu_C = 3(1 -_2\mu_A -_2\mu_C),$$

with $\Gamma_{_1\mu,\Sigma(3)}$ of index dividing 4 in $\Gamma_{_2\mu,\Sigma(3)}$ modulo $PGL(3, \mathbf{C})$-conjugacy by the Remark at the end of 7.21. A computation left to the reader shows that related $_1\mu$ and $_2\mu$ are given by the formula of 11.22.

Proof of (ii) Recalling the notation of 11.14, we have the group $A(4) \approx C_3 \cdot (\mathbf{Z}_2 + \mathbf{Z}_2)$ acting on the simply connected quadratic cone R_{12} with $R_4 = R_{12}/C_3$ and $R = R_{12}/A(4)$. Let φ denote the map $R_4 \to R$ and let (2) denote the tangential intersection point of D_{AA} and D_{AB} in R. As in 11.19, let $_2R$ and $_1R$ denote the blow ups of R and R_4 ($\approx R$) respectively. Setting $_0R = R_{12} \times_R {}_2R$, we have $_1R = R_4 \times_R {}_2R$ and $_2R = R \times_R {}_2R$. Consider the action of $A(4)$ on $_0R$ via the action on its first factor R_{12}. Let $_i\varphi$ denote the maps $_0R \to_i R$ ($i = 1, 2$). Then $_1R = {}_0R/C_3$, $_2R = {}_0R/A(4)$. Set $_i\Gamma =_i \Gamma_{\mu,\Sigma(3)}$ ($i = 1, 2$) and identify $_1\Gamma$ with a subgroup of $_2\Gamma$ via 11.22 (i). Set Γ_0 equal to the intersection of all the conjugates of $_1\Gamma$ in $_2\Gamma$.

Let $_0R'$ denote the open dense subset of $_0R$ on which $A(4)$ acts freely, and let $_iR' =_i \varphi(R'_0)$ ($1 = 1, 2$). Choosing coherent base points $*$ in $_iR'$ ($i =$

$0, 1, 2)$, we have by (8.3.3)

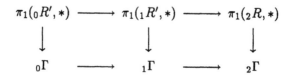

where $_0\Gamma$ is normal in $_2\Gamma$, showing that $_2\Gamma/_0\Gamma$ and $_1\Gamma/_0\Gamma$ are homomorphic images of $A(4)$ and C_3 respectively. Let (AC) denote the blow up in $_2R$ of the point (2) in R. By 11.21, $\varphi^{-1}(AC) =_1 (AC) \amalg (BC)$ where φ is etale on $_1(AC)$ and ramifies along (BC) with index 3; this last assertion follows from the fact that the map φ around BC is equivalent to the map $(T, U) \to (T, U^3)$ along the exceptional curve of 11.20 (b).

Let ξ be an element in $\pi(_1R', *)$ conjugate to a simple loop in $_1R$ around the curve $_1(AC)$. When lifted to $_0R'$, it corresponds to a path from $*$ to the transform of $*$ by a non trivial element of C_3. It follows that ξ is a generator of the cyclic group of order three $\pi(R_1', *)/\pi(R_0', *)$.

Let γ_{AC} denote the image of ξ in $_1\Gamma$. Assume that γ_{AC} has finite order. Referring to the 5-tuple $_1\mu = \mu(\rho, 3, \sigma)$ which defines $_1\Gamma$, the order of γ_{AC} is the numerator of σ. Assume moreover that 3 does not divide the numerator of σ. Then the cyclic group $\langle\gamma_{AC}\rangle$ generated by γ_{AC} coincides with $\langle\gamma_{AC}^3\rangle$. On the other hand ξ^3 comes from a closed loop in $_0R'$ around a lift of $_1(AC)$ in $_0R$, and consequently $\xi^3 \in \pi_1(_0R', *)$. Therefore, $\langle\gamma_{AC}\rangle = \langle\gamma_{AC}^3\rangle \subset {_0\Gamma}$; i.e., $_1\Gamma = {_0\Gamma}$. Denoting normalizers by $N(\ )$, we have $_1\Gamma = N_{_2\Gamma}(_1\Gamma)$ since $N_{A(4)}(C_3) = C_3$. Hence $_1\Gamma = N(_1\Gamma) = N(_0\Gamma) =_2 \Gamma$.

§12. COMMENSURABILITY AND DISCRETENESS

We first summarize the isomorphism theorems of §§10 and 11 in a somewhat more convenient notation. Let

$$(12.1) \qquad \mu = (a, a, b, b, 2 - 2a - 2b)$$

be a 5-tuple of complex numbers. We set

$$(12.1') \qquad \mu' = \left(1 - b, 1 - a, a + b - \frac{1}{2}, a + b - \frac{1}{2}, 1 - a - b\right)$$

and if moreover $a = b$, then we set $\mu'' = (\mu')'$; that is

$$(12.1'') \qquad \mu'' = \left(\frac{3}{2} - 2a, a, a, a, \frac{1}{2} - a\right)$$

We consider the following subgroups of the permutation group $\Sigma(\{1, 2, 3, 4, 5\})$

$$H_1 = \langle (12)(34) \rangle, H_2 \langle (34) \rangle, V = \text{Vierergruppe } \{1, 2, 3, 4\}$$
$$\Sigma(3) = \Sigma(\{1, 2, 3\}), \Sigma(4) = \Sigma(\{1, 2, 3, 4\}) = \Sigma_4, \Sigma_3 = \Sigma(\{2, 3, 4\}).$$

We have by (10.6) and (10.15), modulo conjugacy in $PGL(3, \mathbf{C})$,

$$(12.2) \qquad\qquad \Gamma_{\mu, H_1} = \Gamma_{\mu', H_2}$$

and if $a = b$, then

$$\Gamma_{\mu, V} = \Gamma_{\mu''}$$
$$\Gamma_{\mu, H_1 V} = \Gamma_{\mu'', H_2}$$
$$(12.2') \qquad\qquad \Gamma_{\mu, \Sigma_4} = \Gamma_{\mu'', \Sigma_3}$$

From §11, we get another series of commensurabilities.

If $\pi^{-1} + \rho^{-1} + \sigma^{-1} = 2^{-1}$, set

$$\mu(\pi, \rho, \sigma) = \left(\frac{1}{2} - \frac{1}{\pi}, \frac{1}{2} - \frac{1}{\pi}, \frac{1}{2} - \frac{1}{\pi}, \frac{1}{2} + \frac{1}{\pi} - \frac{1}{\rho}, \frac{1}{2} - \frac{1}{\pi} - \frac{1}{\sigma} \right)$$

If we set $\mu = \mu(3, \rho, \sigma) = \left(\frac{1}{6}, \frac{1}{6}, \frac{1}{6}, \frac{5}{6} - \frac{1}{\rho}, \frac{5}{6} - \frac{1}{\sigma} \right)$, then we set

$$^{\rho}\mu = \mu(\rho, 3, \sigma) = \left(\frac{1}{2} - \frac{1}{\rho}, \frac{1}{2} - \frac{1}{\rho}, \frac{1}{2} - \frac{1}{\rho}, \frac{1}{2} - \frac{1}{\rho}, \frac{1}{6} + \frac{1}{\rho}, 2\left(\frac{1}{6} + \frac{1}{\rho} \right) \right)$$

$$\mu^{\sigma} = \mu(\sigma, 3, \rho).$$

By (11.22)

(12.3) $\Gamma_{^{\rho}\mu, \Sigma(3)} \hookrightarrow \Gamma_{\mu, \Sigma(3)}$

and the image is of finite index.

12.4 We can apply these commensurability theorems to obtain the discreteness of monodromy groups associated to ball 5-tuples which fail to satisfy condition ΣINT of (8.2).

Consider a ball 5-tuple of the form

(12.4.1) $\mu = \left(\frac{1}{2} - \frac{1}{p}, \frac{1}{2} - \frac{1}{p}, \frac{1}{2} - \frac{1}{p}, \frac{1}{2} - \frac{1}{p}, \frac{1}{2} - \frac{1}{p}, \frac{4}{p} \right), \quad p \in \mathbf{Z}.$

Then μ satisfies condition ΣINT if and only if

$$\frac{2p}{p - 6} \left(= \left(1 - \left(\frac{1}{2} - \frac{3}{p} \right) \right)^{-1} \right) \in \mathbf{Z}$$

if positive. By Lemma 8.4.1 and the remark following,

(12.4.2) $p \in \{3, 4, 5, 6, 7, 8, 9, 10, 12, 18, \infty\}.$

For these values of p, $\Gamma_{\mu}, \Gamma_{\mu'}, \Gamma_{\mu''}$ are commensurable by (10.6), provided that $\mu'_i \neq 0, \mu''_i \neq 0$ for all i, and hence are lattice subgroups of $PU(1, 2)$ by (10.8), (10.10) and (8.2).

The 5-tuple

$$\mu' = \left(\frac{1}{2} + \frac{1}{p}, \frac{1}{2} + \frac{1}{p}, \frac{1}{2} - \frac{2}{p}, \frac{1}{2} - \frac{2}{p}, \frac{2}{p} \right)$$

is a ball 5-tuple for $p > 4$, and satisfies condition $\Sigma INT(\{3,4\})$ if and only if p is even. Thus μ' is a non-ΣINT ball 5-tuple with $\Gamma_{\mu'}$ discrete for

$$p = 5 \qquad \mu' = \left(\frac{7}{10}, \frac{7}{10}, \frac{1}{10}, \frac{1}{10}, \frac{4}{10}\right) \qquad (= 60')$$

$$p = 7 \qquad \mu' = \left(\frac{9}{14}, \frac{9}{14}, \frac{3}{14}, \frac{3}{14}, \frac{4}{14}\right) \qquad (= 77')$$

$$p = 9 \qquad \mu' = \left(\frac{11}{18}, \frac{11}{18}, \frac{5}{18}, \frac{5}{19}, \frac{4}{18}\right) \qquad (= 84').$$

Here the numbers inside the parentheses at the right refer to the list in [M], pp. 584-588, of all S-uples satisfying condition ΣINT.

12.5 By Case 1 of (8.3.4), $\Gamma_\mu = \Gamma_{\mu,\Sigma(S_1)}$ if $\mu_s = \frac{1}{2} - \frac{1}{p}$ with p odd for $s \in S_1$. Thus for p odd, we get the isomorphisms, setting $_1\mu = \mu$, $_2\mu = \mu'$, $_3\mu = \mu''$:

$$\Gamma_{1\mu} = \Gamma_{1\mu,\Sigma(\{A_1A_2,A_3,A_4\})} = \Gamma_{3\mu} = \Gamma_{2\mu,\Sigma(\{A_3,A_4\})}.$$

For p even, we get the diagram of inclusions (cf. (10.15.1)), modulo conjugacy in $PU(1,2)$ (cf. 10.18):

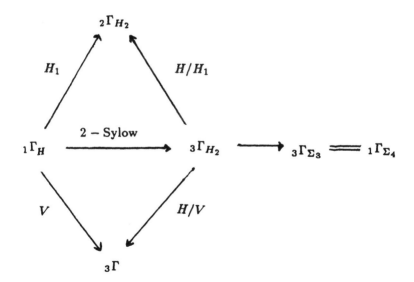

with the indicated quotients, where $_i\Gamma$ denotes $\Gamma_{i\mu}$, $_i\Gamma_H$ denotes $\Gamma_{i\mu,H}$ (cf. (8.3.4) Case 2), $H_1 = \langle (1,2),(3,4)\rangle$, $H_2 = \langle (3,4)\rangle$.

For $4 \mid p$, $_2\mu$ satisfies condition INT and by (8.3.4) Case 2

$$_2\Gamma_{((A_1A_2)(A_3A_4))}/_2\Gamma_{((A_3A_4))} = \mathbf{Z}/2,$$

so that

$$_3\Gamma_{((A_3A_4))}/\mathrm{Im}_2\Gamma_{(A_3A_4)} = \mathbf{Z}/2.$$

Moreover,

$$_2\Gamma_{((A_3A_4))}/_2\Gamma = \mathbf{Z}/2.$$

Finally, we claim that

$$\mathrm{Im}_1\Gamma \text{ in } _2\Gamma_{((A_3A_4))}$$

intersects $_2\Gamma$ in a subgroup of index 2.

PROOF. Set $Q_1 = Q$, $Q_2 = Q \cup D_{C'F_1'} \cup D_{C'F_2'}$, $R = Q_1/H_1 \simeq Q_2/H_2$, and let $\tilde{R}$ denote the largest covering $Q_1/H_1 (\simeq Q_2/H_2)$ described in (8.3.4) on which $\Gamma_{_1\mu H_1} (\simeq \Gamma_{_2\mu H_2})$ acts freely. We have the commutative diagram

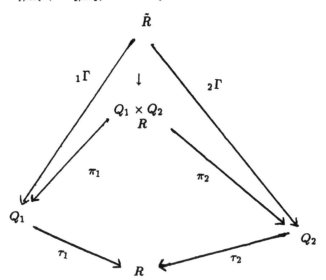

with $\tilde{R}/_i\Gamma = Q_i$ because $_i\mu$ satisfies condition INT, by [DM] (10.18.2). Identifying $\Gamma_{_1\mu,H_1}$ with $\Gamma_{_2\mu,H_2}$, we must show that $_1\Gamma$ does not lie in $_2\Gamma$. Suppose $_1\Gamma \subset_2 \Gamma$. Then the projection $\pi_1 : Q_1 \underset{R}{\times} Q_2 \to Q_1$ would have a continuous section σ, and π_2 would be an even covering. Inasmuch as degree $_2\tau = |H_2| = 2$, $Q_1 \times Q_2$ has two connected components. We would get $_1\tau = \,_2\tau \circ \pi_2 \circ \sigma$ with none of the maps $_2\tau, \pi_2, \sigma$ having ramification. Hence $_1\tau$ would not ramify - a contradiction.

12.6 Another series of non-ΣINT 5-tuples whose monodromy group is discrete arises from (12.3). A necessary and sufficient condition that

$$\mu = \mu(3, \rho, \sigma)$$

be a ball 5-tuple satisfying condition $\Sigma INT(\{1, 2, 3\})$ is that $\rho \in \mathbf{Z}$ if positive, $\sigma \in \mathbf{Z}$ if positive, and $\rho^{-1} + \sigma^{-1} = 6^{-1}$. Moreover, by Lemma 8.4.1 and the Remark following it, either $\rho \geq 2$ or $\rho = -12$ or -30 and similarly $\sigma \geq 2$ or $\sigma = -12$ or -30.

By inspection, $^\rho\mu$ satisfies condition ΣINT if and only if $\rho \in \mathbf{Z}$ and

$$\left(1 - 3\left(\frac{1}{6} + \frac{1}{\rho}\right)\right)^{-1} \in \mathbf{Z}$$

if positive i.e., $\frac{2\rho}{\rho - 6} \in \mathbf{Z}$ or $\frac{\rho - 6}{2\rho} = \frac{1}{n}, n \in \mathbf{Z}$, $\frac{1}{2} - \frac{3}{\rho} = \frac{1}{n}$ or $\frac{1}{\rho} + \frac{1}{3n} = \frac{1}{6}$; that is, $\sigma = 3n$. Therefore $^\rho\mu$ satisfies condition ΣINT if and only if $3 \mid \sigma$ if $\sigma > 0$ (if $\sigma < 0$, then $\sigma = -12$ or -30 and hence $3 \mid \sigma$ automatically).

The equation $\rho^{-1} + \sigma^{-1} = 6^{-1}$ with $\rho, \sigma \in \mathbf{Z}$ has solutions

$$\left\{\begin{matrix} \rho & : & 3 & 4 & 5 & 6 & 7 & 8 & 9 & 10 & 12 & 15 & \cdots \\ \sigma & : & -6 & -12 & -30 & \infty & 42 & 24 & 18 & 15 & 12 & 10 & \cdots \end{matrix}\right\}$$

Thus $^\rho\mu$ is non-ΣINT for $\rho = 15, 24, 42, -30, -12$

$$^{15}\mu = \left(\frac{13}{30}, \frac{13}{30}, \frac{13}{30}, \frac{7}{30}, \frac{14}{30}\right) =^{15} 91$$

$$^{24}\mu = \left(\frac{11}{24}, \frac{11}{24}, \frac{11}{24}, \frac{5}{24}, \frac{10}{24}\right) =^{24} 88$$

$$^{42}\mu = \left(\frac{10}{21}, \frac{10}{21}, \frac{10}{21}, \frac{4}{21}, \frac{8}{21}\right) =^{42} 93$$

$$^{-30}\mu = \left(\frac{8}{15}, \frac{8}{15}, \frac{8}{15}, \frac{3}{15}, \frac{6}{15}\right) =^{-30} 90$$

$$^{-12}\mu = \left(\frac{7}{12}, \frac{7}{12}, \frac{7}{12}, \frac{1}{12}, \frac{2}{12}\right) =^{-12} 63$$

For these ball 5-tuples μ, Γ_μ is discrete but μ is non-ΣINT.

Comparing these eight 5-tuples of 12.4 and 12.6 with the list in [M4] §5.1 of lattices Γ_μ in $PU(1, 2)$ with μ a non-ΣINT ball 5-tuple, we find that we have them all except for

$$\left(\frac{1}{12}, \frac{3}{12}, \frac{5}{12}, \frac{5}{12}, \frac{10}{12}\right)$$

(which is seen to be arithmetic by the criterion of [DM]).

§13. AN EXAMPLE

We thank R. Askey for an illuminating letter explaining to us how classical binomial coefficient identities can profitably be understood as computing the value at $z = 1$ or -1 of some hypergeometric series $_{p+1}F_p$ (cf. [Askey]).
13.1. The aim of this section is to explain how to obtain power series expansions for some branches of Lauricella's hypergeometric functions and how to extract identities between power series from the results of Section 10. We will work out one case completely, making 10.21 explicit. In that case, it eventually boils down to the known identity (Pfaff 1797) expressing the value at 1 of a balanced terminating hypergeometric series $_3F_2$:

$$_3F_2\left(\begin{matrix} -n, a, b \\ c, d \end{matrix}; 1\right) = (-1)^n \frac{(c-a)_n(c-b)_n}{(c)_n(d)_n}$$

(13.1.1)   when $(c+d) - (a+b-n) = 1$.

The latter condition implies that $(c-a)_n = (-1)^n(d-b)_n;\ (c-b)_n = (-1)^n(d-a)_n$ so that the second member has the required symmetry in c and d. We recall that

$$_3F_2\left(\begin{matrix} a, b, c \\ d, e \end{matrix}; z\right) = \sum_n \frac{(a)_n(b)_n(c)_n}{(d)_n(e)_n} \frac{z^n}{n!},$$

where $(a)_n = a(a+1)\dots(a+n-1)$.

The identity (13.1.1) can be obtained by equating coefficients in

(13.1.2) $(1-x)^{a+b-c}\, _2F_1\left(\begin{matrix} a, b \\ c \end{matrix}; x\right) = \, _2F_1\left(\begin{matrix} c-a, c-b \\ c \end{matrix}; x\right)$

13.2 With the notations of §1.N except for the meaning of S_1, assume $N \geq 4$ and let $S = \{S_0, S_1, S_\infty\}$ be a partition of S. We will consider S-uples x of points of $\mathbf{P}^1$ where the three sets $\{x_s \mid \{s \in S_i\}\}$ cluster around three distinct points. Here is the main computation.

We fix disjoint discs $D_0 : |z| < \varepsilon, D_1 = |z - 1| < \varepsilon, D_\infty = |z^{-1}| < \varepsilon$ around $0, 1$ and ∞. If $x_s \in D_i$ whenever $s \in S_i$, we consider the multivalued integrand

$$(13.2.1) \qquad \omega := \prod_{s \in S_0} (t - x_s)^{-\mu_s} \prod_{s \in S_1} (x_s - t)^{-\mu_s} \prod_{s \in S_\infty} (1 - tx_s^{-1})^{-\mu_s} dt.$$

This integrand has a principal determination for t near $[0, 1] - D_0 - D_1$. Set

$$(13.2.2) \qquad \mu[i] := \sum_{s \in S_i} \mu_s.$$

We assume that none of the $\mu[i]$ is an integer. This allows us to define $\alpha_i := \exp(2\pi\sqrt{-1}\mu[i])$ and $\lambda_i := (1 - \alpha_i^{-1})^{-1}$. We also define $\alpha_i^{1/2} := \exp(\pi\sqrt{-1}\mu[i])$.

The integrand (13.2.1) will be integrated on the following cycle Z: a sum of oriented arcs, each provided with a constant multiple of a determination of ω. In the picture, each arc is labelled by which multiple of the principal determination is to be taken. The cycle condition is met (cf. 4.14)

$(13.2.3) \quad z:$

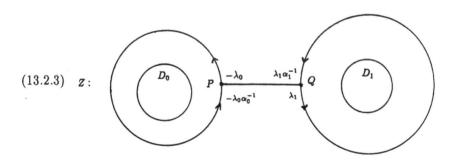

The cycle Z represents a homology class of $\mathbf{P}^1 - D_0 - D_1 - D_\infty$ with coefficients in the local system of constant multiples of ω. The corresponding H^1 is of dimension 1. The integral $\int_Z \omega$ is by definition the sum of

$$\int_P^Q \text{(principal determination of } \omega\text{)},$$

of an integral from P to P around D_0 and an integral from Q to Q around D_1, of a multiple of ω which begins by being $-\lambda_0$ (principal determination) (resp. λ_1 (principal determination)). The cycle conditions read

$$\lambda_0 - \lambda_0 \alpha_0^{-1} - 1 = 0 \quad \text{and}$$
$$- \lambda_1 + \lambda_1 \alpha_1^{-1} + 1 = 0.$$

The integral is well defined and holomorphic in the x_s as long as each x_s remains in its respective disc D_i - whatever their coalescences.

We define new variables y_s by $y_s = x_s$ $(s \in S_0)$, $y_s = 1 - x_s$ $(s \in S_1)$ and $y_s = x_s^{-1}$ $(s \in S_\infty)$. The factors in (13.2.1) can be expanded as follows:

$$s \in S_0 : (t - x_s)^{-\mu_s} = t^{-\mu_s}(1 - x_s/t)^{-\mu_s} = t^{-\mu_s}\Sigma(\mu_s)_n y_s^n t^{-n}/n!$$

$$s \in S_1 : (x_s - t)^{-\mu_s} = (1 - t)^{-\mu_s}(1 - \frac{1 - x_s}{1 - t})^{-\mu_s}$$
$$= (1 - t)^{-\mu_s}\Sigma(\mu_s)_n y_s^n (1 - t)^{-n}/n!$$

$$s \in S_\infty : (1 - tx_s^{-1})^{-\mu_s} = \Sigma(\mu_s)_n y_s^n t^n/n!$$

for t outside of the respective discs D_i.

Taking the product, we obtain a power series expansion of ω in the y_s. It is valid outside of the D_i, and in particular on the integration cycle Z. If we put $\mathbf{n} = (n_s)_{s \in S}$, $|n| = \Sigma n_s$, $|n|_i = \sum_{s \in S_i} n_s$, and $\mathbf{y^n} = \prod_s y_s^{n_s}$ it is

$$(13.2.4) \qquad \omega = dt \sum_{\mathbf{n}} \mathbf{y^n} \frac{\prod(\mu_s)_{n_s}}{\prod n_s!} \cdot t^{-\mu[0] - |n|_0 + |n|_\infty}(1 - t)^{-\mu[1] - |n|_1}.$$

We now integrate $\int_Z \omega$ term by term. As we have shown in [DM], 2.17, if $A \equiv -\mu[0] \bmod \mathbf{Z}$ and $B \equiv -\mu[1] \bmod \mathbf{Z}$, then

$$\int_Z t^A(1 - t)^B dt = Pf \int_0^1 t^A(1 - t)^B dt = \frac{\Gamma(A + 1)\Gamma(B + 1)}{\Gamma(A + B + 2)}.$$

Taking $A \equiv \mu[0] \bmod \mathbf{Z}$ and $B \equiv -\mu[1] \bmod \mathbf{Z}$, we get

$$(13.2.5) \quad \int_Z \omega = \sum_{\mathbf{n}} \mathbf{y^n} \frac{\Gamma(-\mu[0] - |n|_0 + |n|_\infty + 1)\Gamma(-\mu[1] - |n|_1 + 1)}{\Gamma(2 - \mu[0] - \mu[1] - |n|_0 - |n|_1 + |n|_\infty)}.$$

One has $2 - \mu[0] - \mu[1] = \mu[\infty]$, $\Gamma(s + n) = \Gamma(s) \cdot (s)_n$ and $(s)_n = (-1)^n/(1 - s)_{-n}$: (13.2.5) can be rewritten

(13.2.6)

$$\int_Z \omega = C \cdot \sum_\mathbf{n} \mathbf{y}^\mathbf{n} \cdot (-1)^{|\mathbf{n}|}$$

$$\cdot \frac{\prod (\mu_s)_{n_s}}{\prod n_s!} (-\mu[0] + 1)_{-|\mathbf{n}|_0 + |\mathbf{n}|_\infty} (-\mu[\infty] + 1)_{|\mathbf{n}|_0 + |\mathbf{n}|_1 - |\mathbf{n}|_\infty}$$

$$\cdot (-\mu[1] + 1)_{-|\mathbf{n}|_1}$$

with

$$C = \frac{\Gamma(-\mu[0] + 1)\Gamma(-\mu[1] + 1)}{\Gamma(\mu[\infty])}.$$

If we fix $a \in S_0, b \in S_1$ and $c \in S_\infty$ and consider only x with $x_a = 0, x_b = 1$ and $x_c = \infty$, we have $y_a = y_b = y_c = 0$, and (13.2.6) is an expansion for an hypergeometric-like function in $L(\alpha)$ (4.10). Here α is characterized by (cf. Remark following 5.4).

$$\sum_t \alpha_{s,t} = \mu_s$$

$$\alpha_{a,s} = \mu_s \text{ for } s \in S_\infty - \{c\}$$

(13.2.7)

$$\alpha_{s,t} = 0! \text{otherwise, unless } \{s,t\} = \{a,b\}, \{a,c\}, \{b,c\}$$

since the integrand ω can indeed be rewritten

$$(-1)^{\mu[1]} \prod_{s \in S_\infty - \{c\}} (-x_s)^{\mu_s} \prod_{s \neq c} (t - x_s)^{-\mu_s}.$$

In particular,

(13.2.8) $\alpha_{s,t} = 0$ for all $s, t \in S_i, s \neq t$ $(i = 1, 2, 3)$

13.3 Let $M^* \subset \mathbf{P}^{1^S}$ be the subset of x such that $x_s \neq x_t$ if s and t are in different cosets of the partition S of S. Define $Q^* = M^*/PGL(2)$ and let $q \in Q^*$ be the $PGL(2)$-coset of x for which $x_s = x_t$ when s and t are in the same coset of S.

Fix exponents $\alpha_{s,t}$ as in 4.10 and assume that the corresponding $\mu_s := \sum_t \alpha_{s,t}$ are not integers and that $\mu[i]$ is not an integer for $i = 0, 1, \infty$.

The hypergeometric-like local system of holomorphic functions $L(\alpha)$ on Q is obtained by considering integrals of a suitable multivalued relative differential form on $Q(S_+)$, the universal punctured projective line over Q. Let $L^\vee$ be the rank one local system of relative 1-forms spanned by the determinations of ω. Integration of ω over cycles identifies the local system $L(\alpha)$ with the local system whose value at $x \in Q$ is the homology H_1 of the fiber F_x at x of $Q(S_+) \to Q(S)$, with coefficients in $L^\vee$. The fiber F_x is a projective line with $|S|$ punctures. For x in a small neighborhood of q, the punctures cluster in three groups, contained in disjoint discs D_0, D_1, D_∞. The homology

$$H_1(F_x - D_0 - D_1 - D_\infty, L^\vee)$$

is rank one. Its image in $H_1(F_x, L^\vee)$ determines a rank one sub-local system $L_S(\alpha)$, defined in the trace on Q of a neighborhood U of $q \in Q^*$. $L_S(\alpha)$ consists of the functions obtained by integrating ω on a cycle chosen as in 13.2.3. The formulas (13.2.5) (13.2.6) give power series expansions near q for functions in $L_S(\alpha)$.

Fix $s, t \in S$, in the same coset of $\mathcal{S}$, with $\mu_s + \mu_t \notin \mathbf{Z}$.

Along $D_{s,t}$, the local system $L(\alpha)$ is a direct sum $L'(\alpha) \oplus L''(\alpha)$, with $L''(\alpha)$ the $\exp(2\pi i\alpha_{s,t})$-eigenspace of the monodromy around $D_{s,t}$ and of codimension 1 (cf. 6.7). Moreover, $L_S(\alpha)$ inherits its monodromy around $D_{s,t}$ from $L^\vee$. Therefore, by 13.2.8, one has

$$L(\alpha)_S \subset L''(\alpha).$$

We show next that if the $\mu_s + \mu_t$ are not in $\mathbf{Z}$, this property characterizes $L(\alpha)_S$.

PROPOSITION 13.4. *Assume that $\mu_s \notin \mathbf{Z}$, $\mu[i] \notin \mathbf{Z}$ $(i = 0, 1, \infty)$ and $\mu_s + \mu_t \notin \mathbf{Z}$ for s, t in the same coset of $\mathcal{S}$. Assume $U \cap Q$ connected (this can be achieved by shrinking U). If $L_1 \subset L(\alpha)$ is a nonzero sublocal system of $L(\alpha)$ on $U \cap Q$, such that for each s, t in the same coset of $\mathcal{S}$, one has $L_1 \subset L''(\alpha)$ along $D_{s,t}$, then $L_1 = L_S(\alpha)$.*

PROOF. To prove the proposition, one is free to shrink U. In particular, one can choose U as follows: fix $a \in S_0, b \in S_1, c \in S_\infty$, fix open discs D_0, D_1, D_∞ around $0, 1, \infty \in \mathbf{P}^1$, identify Q with the space of injective maps x from S to $\mathbf{P}^1$ such that $x(a) = 0, x(b) = 1, x(c) = \infty$, and define U by the conditions $x(s) \in D_i$ for $s \in S_i$.

Fix a base point $x_0 \in U$. The fiber of $L(\alpha)$ at x is $H_1(F_x, L^{\vee})$.

Let bD_i be the circle $\bar{D}_i - D_i$. The complement in F_x of the bD_i decomposes into $F_x - U\bar{D}_i$ and the $D_i - x(S_i)$. As $\mu[i] \notin \mathbf{Z}$, one has $H^*(bD_i, L^{\vee}) = 0$, and the long exact sequence of cohomology for F_x and the bD_i reduces to a direct sum decomposition

$$H_c^*(F_x - U\bar{D}_i, L^{\vee}) \oplus \bigoplus_i H_c^*(D_i - x(S_i), L^{\vee}) \xrightarrow{\sim} H_c^*(F_x, L^{\vee}).$$

The fiber of $L(\alpha)_S$ at x is the first summand.

The local system $L^{\vee}$ is the dual of L, and $H_c^*(D_i - x(S_i), L^{\vee})$ is dual to $H^*(D_i - x(S_i), L)$. In D_i, embed a tree T_i having as vertices the points of $x(S_i)$. For each edge e, choose an orientation or of e and a nonzero section ℓ of L on e. Then, (e, or, ℓ) defines a class in $H^*(D_i - x(S_i), L)$. An argument similar to the one given in [DM] (2.5.1), using $\mu[i] \notin \mathbf{Z}$, shows that such classes form a basis of $H^*(D_i - x(S_i), L)$. The fiber of $L(\alpha)_S$ at x is hence the orthogonal of all such classes (e, or, ℓ) for $i = 0, 1, \infty$.

Let e be any edge of T_i, going from $x(s)$ to $x(t)$. As we move $x(s)$ toward $x(t)$ along e, the point x moves in U and approaches $D_{s,t}$, so that $L''(\alpha)$ becomes defined. If we prove that it is the orthogonal of (e, or, ℓ), the assumption $L_1 \subset L''(\alpha)$ means that L_1 is orthogonal to (e, or, ℓ) and the proposition follows.

That $L''(\alpha)$ is the orthogonal of (e, or, ℓ) is a local question in the neighborhood of a point of $D_{s,t}$. We now choose x in such a neighborhood. We may assume that $x(s), x(t)$ and e are contained in a small disc D not containing any other $x(i)$. Because $\mu_s + \mu_t \notin \mathbf{Z}$, one has $H^*(bD, L) = 0$ and again a direct sum decomposition

$$H_c^1(D - x(s) - x(t), L^{\vee}) \oplus H_c^1(F_x - D, L^{\vee}) \xrightarrow{\sim} H_c^1(F_x, L^{\vee}).$$

By matching monodromy, we see that it is the decomposition (cf. 13.2.8) $L'(\alpha) \oplus L''(\alpha) = L(\alpha)$. One has a dual decomposition

$$H^1(D - x(s) - x(t), L) \oplus H^1(F_x - D, L) \xleftarrow{\sim} H^1(F_x, L).$$

The class (e, or, ℓ) is a basis of $H^1(D - x(s) - x(t), L)$ and is orthogonal to $H_c^1(F_x - D, L^{\vee})$. This completes te proof.

13.5 Let us consider the space Q_2 of configurations (mod $PGL(2)$) of 5-uples (F_1, F_2, A, B, C). We fix $F_1 = 0, F_2 = \infty, C = 1$ and use as coordinates on Q_2 the coordinate of A and the inverse of the coordinate of B.

In these coordinates, the interchange of F_1 and F_2 is the map

$$(x, y) \mapsto (x^{-1}, y^{-1}).$$

As in 10.21, we choose exponents $_2\alpha$, invariant by the permutation of F_1 and F_2 and such that $_2\alpha_{F,C} = 0$, $_2\beta_{F,C} = \frac{1}{2}$. If we twist by functions $[(1-x)(1-x^{-1})]^A, [(1-y)(1-y^{-1})]^B$, we can assure $_2\alpha_{AF} = {}_2\alpha_{BF} = 0$. Since $1 - {}_2\mu_F - {}_2\mu_C = \frac{1}{2}$ the α's are then determined by $_2\mu_A$ and $_2\mu_B$; they are $_2\alpha_{AB} = \frac{1}{2}$, $_2\alpha_{AC} = {}_2\mu_A - \frac{1}{2}$, $_2\alpha_{BC} = {}_2\mu_B - \frac{1}{2}$.

The multivalued relative 1-form on $Q(S^+)$ to be integrated is then writing μ_A for $_2\mu_A$ and μ_B for $_2\mu_B$,

(13.5.1)

$$\omega = (1 - xy)^{1/2}(1 - x)^{\mu_A - \frac{1}{2}}(1 - y)^{\mu_B - \frac{1}{2}}$$
$$\cdot t^{-(\frac{3}{2} - \mu_A - \mu_B)}(1 - t)^{-(\mu_A + \mu_B - 1)}(t - x)^{-\mu_A}(1 - yt)^{-\mu_B} dt.$$

13.6 Let us consider the space Q_1 of configurations (mod $PGL(2)$) of 5-tuples (A_1, A_2, B_1, B_2, C). Via the birational isomorphism

$$Q_1/\mathbf{Z}(2) \times \mathbf{Z}(2) \to Q_2/\mathbf{Z}(2)$$

of 10.3, the point of Q_2 with coordinates (x, y) corresponds to the 5-uple $(\pm\sqrt{x}, \pm\sqrt{y}^{-1}, 1)$.

We will use (x, y) as coordinates on Q_1. It might be better to say: as local coordinates on Q_1, since the functions x and y are defined only on the double covering of Q_1 which parametrizes configurations (A_1, A_2, B_1, B_2, C) together with an ordering (F_1, F_2) of the two fixed points of the involution for which $(A_1, A_2), (B_1, B_2)$ are in involution.

The exponents $_1\alpha$ corresponding by 10.21 to $_2\alpha$ correspond to the hypergeometric-like local system obtained by taking integrals of the multivalued relative 1-form η on the universal punctured projective line over Q_1, given by

$$\eta = (1 - xy)^{\mu_A - \mu_B + 3/2}(1 - x)^{\mu_A - 1/2}(1 - y)^{\mu_B - 1/2}$$

(13.6.1)

$$\cdot (t^2 - x)^{-(1 - \mu_B)}(1 - yt^2)^{-(1 - \mu_A)}(1 - t)^{-2(\mu_A + \mu_B - 1)} dt.$$

This formula makes sense on Q_1, not only on its double covering, because if x, y, t are replaced by $x^{-1}, y^{-1}, t^{-1}, \eta$ remains the same (up to a constant factor depending on the choices of determinations).

We partially compactify Q_2 and Q_1 by allowing $A = F_1, B = F_2$ for Q_2 and $A_1 = A_2, B_1 = B_2$ for Q_1. These partial compactifications correspond by 10.7. If we use cycles Z as in 13.2 for the partitions $\{\{A, F_1'\}, \{C'\},$ $\{B, F_2'\}\}$ and $\{\{A_1, A_2\}, \{C\},$ $\{B_1, B_2\}\}$, Theorem 10.21 (2) and the discussion 13.3 give a proportionality between $\int_{Z_2} \omega$ and $\int_{Z_1} \eta$: for x and y close to 0,

$$(1 - xy)^{1/2}(1 - x)^{\mu_A - 1/2}(1 - y)^{\mu_B - 1/2}$$

$$\cdot \int_{Z_2} t^{-(\frac{3}{2} - \mu_A - \mu_B)}(1 - t)^{-(\mu_A + \mu_B - 1)}(t - x)^{-\mu_A}(1 - yt)^{-\mu_B} dt$$

$$= \lambda \cdot (1 - xy)^{-\mu_A - \mu_B + 3/2}(1 - x)^{\mu_A - 1/2}(1 - y)^{\mu_B - 1/2}$$

$$\cdot \int_{Z_1} (t^2 - x)^{-(1 - \mu_B)}(1 - yt^2)^{-(1 - \mu_A)}(1 - t)^{-2(\mu_A + \mu_B - 1)} dt.$$

The proportionality constant λ is computed by making $x = y = 0$:

$$(13.6.3) \qquad \frac{\Gamma(\mu_B - \frac{1}{2})\Gamma(2 - \mu_A - \mu_B)}{\Gamma(\frac{3}{2} - \mu_A)} = \lambda \frac{\Gamma(2\mu_B - 1)\Gamma(3 - 2\mu_A - 2\mu_B)}{\Gamma(2 - 2\mu_A)}.$$

By the duplication formula

$$\Gamma(s) = (2\pi)^{1/2} 2^{s - 1/2} \Gamma(\frac{s}{2})\Gamma(\frac{s}{2} + \frac{1}{2}),$$

this amounts to

$$(13.6.4) \qquad \lambda = s^{1/2}(2\pi)^{1/2}\Gamma(1 - \mu_s)/\Gamma(\mu_B)\Gamma(\frac{3}{2} - \mu_s - \mu_B)$$

13.7 We simplify notations by putting $\mu := \mu_A$, $\nu := \mu_B$. If we expand the integrals in (13.6.2) as in 13.2 and normalize the resulting power series to have constant term 1, we obtain the following identity:

(13.7.1)

$$(1 - xy)^{1/2}(1 - x)^{\mu - \frac{1}{2}}(1 - y)^{\nu - \frac{1}{2}}$$

$$\sum_{k,\ell} \frac{x^k}{k!} \frac{y^\ell}{\ell!}(-1)^{k+\ell}(\mu)_k(\nu)_\ell(\mu - 1/2)_{k-\ell}(\nu - 1/2)_{\ell-k}$$

$$= (1 - xy)^{3/2 - \mu - \nu}(1 - x)^{\mu - 1/2}(1 - y)^{\nu - 1/2}$$

$$\cdot \sum_{k,\ell} \frac{x^k}{k!} \frac{y^\ell}{\ell!}(1 - \nu)_k(1 - \mu)_\ell(2\mu - 1)_{2(k-\ell)}(2\nu - 1)_{2(\ell-k)}.$$

Let us unravel identity (13.7.1). If we divide both sides by

$$(1 - xy)^{1/2}(1 - x)^{\mu-1/2}(1 - y)^{\nu-1/2},$$

it becomes

(13.7.2)

$$\Sigma \frac{x^k y^\ell}{k! \ell!}(-1)^{k+\ell}(\mu)_k(\nu)_\ell(\mu - 1/2)_{k-\ell}(\nu - 1/2)_{\ell-k} =$$

$$(1 - xy)^{1-\mu-\nu}\Sigma\frac{x^k y^\ell}{k! \ell!}(1 - \nu)_k(1 - \mu)_k(1 - \mu)_\ell(2\mu - 1)_{2(k-\ell)}(2\nu - 1)_{2(\ell-k)}.$$

Let T be the operation which to a power series $\Sigma c_{k,\ell} x^k y^\ell$ attaches the power series $\Sigma c'_{k,\ell} x^k y^\ell$ with

$$c'_{k,\ell} = c_{k,\ell}/(-1)^{k-\ell}(\mu - 1/2)_{k-\ell}(\nu - 1/2)_{\ell-k}.$$

As the quantity by which we divide depends only on $k - \ell$, T commutes with a multiplication by a power series of the form $g(xy)$. One has for $A \in \mathbf{Z}$

$$(2\mu - 1)_{2A} = 2^{2A}(\mu - 1/2)_A(\mu)_A.$$

The transformation T hence transforms (13.7.2) into

$$(1 - x)^{-\mu}(1 - y)^{-\nu}$$

(13.7.3)

$$= (1 - xy)^{1-\mu-\nu}\Sigma\frac{x^k y^\ell}{k! \ell!}(-1)^{k+\ell}(1 - \nu)_k(1 - \mu)_\ell(\mu)_{k-\ell}(\nu)_{\ell-k}.$$

One has $(-1)^\ell(1 - \mu)_\ell = (\mu - \ell)_\ell$, $(\mu - \ell)_\ell(\mu)_{k-\ell} = (\mu - \ell)_k$, similarly for ν, and (13.7.3) can be rewritten

(13.7.4) $(1 - x)^{-\mu}(1 - y)^{-\nu} = (1 - xy)^{1-\mu-\nu}\Sigma\dfrac{x^k y^\ell}{k! \ell!}(\mu - \ell)_k(\nu - \ell)_\ell.$

Equating term by term, we get

(13.7.5) $\dfrac{(\mu)_k}{k!}\dfrac{(\nu)_\ell}{\ell!} = \displaystyle\sum_i (\mu+\nu-1)_i(\mu+i-\ell)_{k-i}(\nu+i-k)_{\ell-i}/i!(k-i)!(\ell-i)!.$

We rewrite the factors to the right

$$(\mu_i - \ell)_{k-i} = (\mu - \ell)_k/(\mu - \ell)_i$$
$$(\nu + i - k)_{\ell-i} = (\nu - k)_\ell/(\nu - k)_i$$
$$(k - i)! = (-1)^i k!/(-k)_i$$
$$(\ell - i)! = (-1)^i \ell!/(-\ell)_i,$$

to transform (13.7.5) into

$$(13.7.6) \qquad \sum_i \frac{(\mu+\nu-1)_i(-k)_i(-\ell)_i}{(\mu-\ell)_i(\nu-k)_i i!} = \frac{(\mu)_k(\nu)_\ell}{(\mu-\ell)_k(\nu-k)_\ell}.$$

The left side is the terminating balanced hypergeometric series

$${}_3F_2\left(\begin{matrix} -k,\mu+\nu-1,-\ell \\ \mu-\ell,\nu-k \end{matrix};1\right).$$

One has $(\nu)_\ell/(\nu-k)_\ell = (\nu-k+\ell)_k/(\nu-k)_k = (-1)^k(-\nu-\ell+1)_k/(\nu-k)_k$: the right side can be rewritten

$$(-1)^k\frac{(\mu)_k(-\nu-k+1)_k}{(\mu-\ell)_k(\nu-k)_k}$$

and 13.7.6 is, as promised, a special case of Pfaff identity (13.1.1): the case when b is a negative integer.

If one wanted to, one could reverse the computation in this paragraph and deduce (13.7.1) from (13.1.1).

13.8 There are other cases where a three part partition of $\{A_1, A_2, B_1, B_2, C\}$ corresponds to a three part partition of $\{A, B, F_1, F_2, C\}$ so that corresponding hypergeometric functions can be matched as in 13.6. Here is a list:

a) cases where the corresponding spaces Q^* are open subsets of $\bar{Q}_1$ and $\bar{Q}_2$:

(1)	A_1A_2	B_1B_2	C	:	AF_1	BF_2'	C
(2)	A_1B_1	A_2B_2	C	:	F_1F_2C	A	B
(3)	A_1A_2	B_1C	B_2	:	AF_1	BC	F_2
(4)	A_1B_1	A_2C	B_2	:	F_1F_2	AC	B

and the cases deduced by permuting the A's, B's, F's or the role of A and B

b) cases where corresponding lines in $\bar{Q}_1$ and $\bar{Q}_2$ should first be blown down:

(5)	A_1A_2B	B_2	C	:	AF_1F_2	B	C
(6)	A_1A_2C	B_1	B_2	:	AF_2C	F_1	B

(for (5), contract $C = B_1, C = B_2$, and $C = B$; for (6), contract $B_1 = B_2$ and $B = F_1, B = F_2$) and cases deduced by symmetry as before.

Case (1) is the one we have unraveled. Here are the results for case (5).

13.9 On Q_2, we use local coordinates (a, b) with

$$(A, B, F_1, F_2, C) = (a, b, 0, \infty, 1).$$

The corresponding point on Q_1 is given by

$$(A_1 A_2 B_1 B_2 C) = (\sqrt{a}, -\sqrt{a}, \sqrt{b}, -\sqrt{b}, 1).$$

We take b close to 1, a away from 1 and define $\sqrt{b}$ to be the square root of b close to 1. We put

$\omega = (b-a)^{1/2}(1-a)^{\mu_a-1/2}(b-1)^{\mu_b-1/2}z^{-(\frac{3}{2}-\mu_a-\mu_b)}(z-1)^{-(\mu_a+\mu_b-1)}$

$(z-a)^{-\mu_a}(b-z)^{-\mu_b}dz$

$\eta = (1-a)^{\mu_a-1/2}(b-1)^{\mu_b-1/2}(b-a)^{-\mu_a-\mu_b+\frac{3}{2}}$

$(z^2-a)^{-(1-\mu_b)}(b-z^2)^{-(1-\mu_a)}(z-1)^{-2(\mu_a+\mu_b-1)}dz$

and obtain for some constant c

$$\int_1^{\sqrt{b}} \eta = c \int_1^b \omega.$$

The constant can be computed by taking the limit for $b \to 1$.

§14. ORBIFOLD

It will be convenient for us to use the language of *orbifolds*, a notion which has appeared in the literature with various side conditions and called by other names: the *V-manifolds* of Satake, the *algebraic stacks* of Grothendieck, and indeed the *orbifolds* of Thurston.

The variant we will use is that of orbifolds which are locally the quotient of a non singular analytic variety by a finite group acting faithfully. In this particular case, the notion of orbifold can be presented as follows.

14.1. DEFINITION. *An orbifold* purely of dimension n is a normal analytic space M, purely of dimension n, together with a locally finite collection $\mathcal{D}$ of irreducible Weil divisors (i.e. codimension one cycles) D the *ramification divisors*, to each of which is assigned a *weight* $\lambda(D)$, the *ramification weight*, of the form $1/k$, where k, the *ramification index*, is an integer at least two.

Moreover, M is covered by the image of *orbifold charts*, that is morphisms $\varphi : V \to M$ where
(i) V is an open subset of $\mathbf{C}^n$;
(ii) the fibers of φ are discrete in V;
(iii) after deletion from V of some subspace Σ of V of codimension at least two, the map φ is etale outside of the $\varphi^{-1}(D)(D \in \mathcal{D})$ and above D, ramifies with ramification index $\lambda(D)^{-1}$.

To an irreducible divisor E not in $\mathcal{D}$, we assign the weight $\lambda(E) = 1$.

The *induced* orbifold structure on an open subset U has as ramification divisors the irreducible components of the $U \cap D(D \in \mathcal{D})$, with weight $\lambda(D)$; there is a similar definition, for $u : U \to M$, U etale over M, with $U \cap D$ replaced by $u^{-1}(D)$.

14.2. **Remarks** (i) It follows from condition 14.1 (ii) that for $y \in V$, with image x in M, there are open neighborhoods V' of y and M' of x such that

φ induces a map $\varphi' : V' \to M'$ which is finite (i.e. proper with finite fibers). The image $\varphi(V')$ is then a closed subspace of M', of the same dimension as M'. As M' is normal, $\varphi(V')$ contains a neighborhood of x. It follows that (a) the map φ is open, and (b) for M' a small enough open neighborhood of x, the connected component V_1 of $\varphi^{-1}(M')$ containing y is finite over M'. For M' small enough, $\varphi^{-1}(x) \cap V_1$ is then reduced to $\{y\}$.

(ii) In 14.1 (iii) we may and shall assume that Σ contains the inverse image of the singular set of M as well as the inverse image of the singular set of the union $\cup \mathcal{D}$ of the ramification divisors: both are of codimension at least two. The condition 14.1 (iii) concerns the map φ' induced by φ from $V' := V - \Sigma$ to M: it means that at $y \in V'$ with image $x \in D$, in suitable local coordinates $(z_1, \ldots z_n)$ of V and M centered respectively at y and x the map φ' is given by $(z_1, z_2, \ldots, z_n) \to (z_1^k, z_2, \ldots, z_n)$, with $k = \lambda(D)^{-1}$ and with $z_1 = 0$ an equation for D.

(iii) Local charts are locally unique, in the following sense:

14.3. PROPOSITION. *For $i = 1, 2$, let $\varphi_i : V_i \to M$ be orbifold charts, and let $y_i \in V_i$ have image x in M. There exist neighborhoods V_i' of y_i in V_i and an isomorphism of V_1' with V_2' for which the following diagram is commutative:*

$$V_1' \qquad \overset{\sim}{\longrightarrow} \qquad V_2'$$

$$\searrow \varphi_1 \qquad \swarrow \varphi_2$$

$$M$$

We first prove

14.4 LEMMA. *Let $\varphi : V \to M$ be an orbifold chart. Let M_1 be normal, of the same dimension as M and let $f : M_1 \to M$ be a morphism. Assume that M can be covered by open sets U, such that the connected components of $f^{-1}(U)$ are finite over U. Assume further that outside of a suitable subspace Σ_1 of M_1, of codimension at least two, the map f is etale outside of the $f^{-1}(D)(D \in \mathcal{D})$ and on $f^{-1}(D)$ ramifies with ramification index dividing $\lambda(D)^{-1}$.*

Then, the sheaf on V of liftings $g : V \to M_1$ of φ is locally constant.

We may and shall assume that Σ_1 contains the singular set $Sing(M_1)$ of M_1, as well as $f^{-1}(Sing\, M)$ and $f^{-1}(Sing(\cup \mathcal{D}))$. As in 14.2 (i), the map

f is open. The condition on $f : M_1 - \Sigma_1 \to M$, above D, is as in 14.2 (ii), but this time with only a divisibility $k \mid \lambda(D)^{-1}$.

PROOF. We may and shall assume that the map f is finite: one replaces in turn M by one of the set of open subsets U assumed to cover M, V by $\varphi^{-1}(U)$, and M_1 by one of the connected components of $f^{-1}(U)$.

As f is finite, enlarging Σ to $\Sigma \cup \varphi^{-1} f \, \Sigma_1$, we may and shall assume that $\Sigma \supset \varphi^{-1} f \, \Sigma_1$.

If Σ and Σ_1 are empty, the claim now results from a direct computation, which can be reduced to the case where φ is $\mathbf{C}^n \to \mathbf{C}^n : (z_1, \ldots, z_n) \to (z_1^k, z_2 \ldots, z_n)$ and f is $\mathbf{C}^n \to \mathbf{C}^n : (z_1, \ldots, z_n) \to (z_1^\ell, z_2, \ldots, z_n)$, with $\ell \mid k$. The local liftings are the maps $(z_1, \ldots, z_n) \to (\alpha z_1^{k/\ell}, z_2 \ldots z_n)$, for α an ℓth roots of unity.

We now treat the general case. It suffices to show that for any $y \in V$, any simply connected open neighborhood V' of y in V, and any $z \in V' - \Sigma$, a local lifting at z extends uniquely to a lifting on V'. The Σ and Σ_1 empty case, applied to $M - f(\Sigma_1)$, $M_1 - f^{-1} f \Sigma_1$ and $V - \Sigma$, shows that the sheaf of local liftings is locally constant on $V - \Sigma$. As Σ is of codimension ≥ 2 in V', which is non singular, $V' - \Sigma$ is simply connected. Local liftings at z hence extend uniquely over $V' - \Sigma$. They extend even to V' because V' is normal and f is finite; for, localizing on M, we may assume that M, hence M_1, embeds into a bounded open subset of $\mathbf{C}^N$. The section s on $V' - \Sigma$ then extends as a map $\bar{s}$ to $\mathbf{C}^N$. On the other hand, let $\Gamma \subset V' \times_M M_1 \subset V \times M_1$ be the graph of s. As M_1 is proper over M, $V' \times_M M_1$ is proper over V', and therefore the closure of $\bar{\Gamma}$ of Γ is proper over V'. The projection of $\bar{\Gamma}$ in V' is closed, contains $V' - \Sigma$, and hence is the whole of V'. The closure $\bar{\Gamma}$ being contained in the graph of $\bar{s}$ and projecting onto V', coincides with the graph of $\bar{s}$. It follows that $\bar{s}$ maps V' into M_1; it is the desired extension.

PROOF OF 14.3. By 14.2 (i), replacing M by a small enough open connected neighborhood M' of x, and V_i by the connected component of $\varphi_i^{-1}(M')$ containing y_i, we may and shall assume that V_i is finite over M and that $\varphi_i^{-1}(x)$ is reduced to y_i; this implies that φ_i^{-1} of a small neighborhood of x is a small neighborhood of y_i. Let now M' be an open neighborhood of x such that $\varphi_i^{-1}(M')$ is contained in a simply connected neighborhood V_i'' of y_i, let V_i' be the connected component of $\varphi_i^{-1}(M')$ containing y_i and choose points $z_i \in V_i'$ at which φ_i is etale, each over the same point of M'. By 14.4 applied to V_1'' and V_2, there is a lifting of φ_1, $u_1 : V_1'' \to V_2$, mapping z_1 to z_2. It induces a lifting $v_1 : V_1' \to V_2'$, mapping z_1 to z_2. Since φ_i is etale at z_i, this lifting is unique near z_1 and by analytic continuation, this lifting is unique. Similarly, we obtain $v_2 : V_2' \to V_1'$, mapping z_2 to z_1. The composite $v_2 v_1$ maps z_1 to itself. By unicity, it is the identity. So also is $v_1 v_2$, and v_1 is the promised isomorphism.

14.5. Let V be an orbifold chart for the orbifold M. Let $y \in V$ have image x in M. The proof of 14.3, applied to $V_1 = V_2 = V$, shows that if M' is a small enough open connected neighborhood of x, and we define V' to be the connected component of $\varphi^{-1}(V)$ containing y, then (a) V' is finite over M': $\varphi(V')$ is then open and closed in M', hence is the whole of M'; (b) $\varphi^{-1}(x)$ is reduced to y; (c) for z in M' above which V is etale and z_1, z_2 above z, V' has a unique automorphism v, compatible with its projection to M, mapping z_1 to z_2. Let Γ be the group of automorphisms of V' compatible with its projection to M. It is finite. The map $V'/\Gamma \to M'$ is finite, open, an isomorphism above a dense open subset of M', and V'/Γ is Hausdorff. It follows that $V'/\Gamma \to M'$ is an homeomorphism. As V'/Γ and M' are normal, it is an isomorphism.

The point y is a fixed point of the action of Γ. In suitable local coordinates centered at y, the action of Γ is linear and one concludes

PROPOSITION 14.6. *For each point x of an orbifold M, there exist a connected open set $V \subset \mathbf{C}^n$ containing 0, stable by a finite group $\Gamma \subset GL(n, \mathbf{C})$ of automorphisms of $\mathbf{C}^n$, and an orbifold chart $\varphi : V \to M$, with $\varphi(0) = x$, which induces an isomorphism of V/Γ with a neighborhood of x in M.*

14.7. Examples. (i) Let Γ be a discrete group acting properly on a non singular analytic space X. We assume that no non trivial element of Γ fixes a non empty open subset of X (this is the case for a faithful action on a connected space). The *orbifold quotient* X/Γ is defined as the space X/Γ together with the collection of images of all the irreducible divisors D in X such that $\Gamma(D) := \{g \in \Gamma; gx = x \text{ for all } x \in D\} \neq \{1\}$, and with weight $\frac{1}{|\Gamma(D)|}$ assigned to the image of D. Local charts of X provide orbifold local charts of X/Γ. It should be remarked that $\Gamma(D)$ is cyclic for any divisor

D, and any of its elements $\neq 1$ may be called a *complex reflection* on X.

An connected orbifold M is called *good* if it is an orbifold quotient X/Γ with X and Γ as above. By 14.6, any orbifold is locally good.

(ii) More generally, let Γ be a discrete group acting properly, by automorphisms, on an orbifold X, again with no non empty open subset fixed by a non trivial element. The *orbifold quotient* X/Γ has X/Γ as underlying space. The ramification divisors are the images of the ramification divisors of X, and of the irreducible divisors with non trivial fixer. The weight assigned to the image of D is $\lambda(D)/|\Gamma(D)|$. Orbifold charts of X provide orbifold charts for X/Γ.

14.8. One defines the notion of orbifold morphism, orbifold universal covering, orbifold fundamental group etc. parallel to the usual notion for complex spaces by the use of orbifold charts.

Let M and N be orbifolds and let $\psi : M \to N$ be a morphism between the underlying normal analytic spaces. We assume for simplicity that ψ is open. In that case, ψ is called an *orbifold morphism* if for all $x \in M$ there is an orbifold chart $\varphi_M : V \to M$ covering x such that $\psi \cdot \varphi_M$ factors through some orbifold chart $\varphi_N : W \to N$ of N:

$$
\begin{array}{ccc}
V & \xrightarrow{\tilde{\psi}} & W \\
{\scriptstyle \varphi_M} \downarrow & & \downarrow {\scriptstyle \varphi_N} \\
M & \xrightarrow{\psi} & N
\end{array}
$$

Shrinking V and W, one may take V connected and W as in 14.6. Two liftings $\tilde{\psi}$ of $\psi\varphi_M$ are then conjugate by an automorphism of W.

A morphism of spaces $\psi : M \to N$ is said to be *orbifold etale* if whenever $\varphi : V \to M$ is an orbifold chart of M, $\psi\varphi$ is an orbifold chart of N. This means that ψ is an orbifold morphism and that any of the above liftings $\tilde{\psi}$ is etale.

An *orbifold covering* is an etale orbifold morphism $\psi : M \to N$ such that N can be covered by open sets U for which the connected components of $\psi^{-1}(U)$ are finite over U.

14.9. Examples. (i) If Γ acts on X as in 14.7 (ii), X is an orbifold covering of the orbifold quotient X/Γ.

(ii) Assume that the orbifolds M and N have the same dimension, and that the map of spaces $\psi : M \to N$ has discrete fibers (cf. 14.2 (i)). It follows immediately from the definition (14.1) of orbifold charts and the definition of orbifold etale that, if ψ is orbifold etale outside a subspace Σ of M of codimension at least two, then ψ is orbifold etale.

(iii) Take for M (resp. N) the affine line. Assume that M and N have no ramification divisor other than 0, and that the weight of 0 is $1/m$ (resp. $1/n$).

The map $z \mapsto z^k$ from M to N is an orbifold morphism if and only if n divides km. It is orbifold etale if and only if $n = km$. It is then an orbifold covering.

14.10. The usual relations between coverings, the universal covering and the fundamental group extend to orbifolds. A connected orbifold M admits a universal orbifold covering $f : \tilde{M} \to M$, that is a connected orbifold covering such that, for any connected orbifold covering $\varphi : M_1 \to M$, f factors through φ. Fix a base point 0 in M, not in $Sing\ M$ nor in $\bigcup \mathcal{D}$. A universal orbifold covering of $(M, 0)$ is a universal covering $\tilde{M}$ of M, given with a base point 0 above that of M. It is unique up to unique isomorphism. More precisely, for any orbifold covering $\varphi : M_1 \to M$ of M and any $z \in \varphi^{-1}(0)$, there is a unique covering map $u : \tilde{M} \to M_1$ with $u(0) = z$.

The orbifold *fundamental group* $\pi_1(M, 0)$ is the automorphism group of the covering $\tilde{M}$ of M, for $(\tilde{M}, 0)$ the universal orbifold covering of M. It acts on the fiber at 0 of any covering M_1 by the following rule: if $u : \tilde{M} \to M_1$ maps 0 to z, then $\gamma z := u \circ \gamma^{-1}(0)$. The functor "fiber above 0" is an equivalence of categories between coverings of M and $\pi_1(M, 0)$-sets.

The orbifold fundamental group can also be described in terms of paths, in the following two equivalent ways:

14.10.1. Each divisor D in $\mathcal{D}$ defines a conjugacy class of elements s_D in the usual fundamental group $\pi_1(M - Sing\ M - \bigcup \mathcal{D}, 0)$; namely, the monodromy around D, near a general (smooth) point of D. If $n(D) := \lambda(D)^{-1}$, $\pi_1(M, 0)$ is the quotient of $\pi_1(M - Sing\ M - \bigcup \mathcal{D}, 0)$ by the relations $s_D^{n(D)} = 1$. More precisely, if U is the covering of $M - Sing\ M - \bigcup \mathcal{D}$ corresponding to that quotient, the universal orbifold covering $\tilde{M}$ is the completion of U over M.

14.10.2. Let us say that two maps $\alpha, \beta : (S^1, 0) \to (M - Sing\ M - \bigcup \mathcal{D}, 0)$ are homotopic in the orbifold M if there is a map $u : S^1 \times [0, 1] \to M$

such that (a) $u(\cdot,0)$ is α and $u(\cdot,1)$ is β; (b) u maps $0 \times [0,1]$ to 0; (c) $u^{-1}(Sing\ M \cup \bigcup \mathcal{D})$ consists of isolated points and at each of them, the map u can be lifted to an orbifold chart of M. The group $\pi_1(M,0)$ can then be identified with the set of homotopy classes of maps of the type considered.

14.11. Remark. If $\tilde{M}$ is a universal orbifold covering of the connected orbifold M, and Γ the group of its automorphisms, compatible with the projection to M, the orbifold M is the orbifold quotient $\tilde{M}/\Gamma$. It follows that M is good if and only if its orbifold universal covering is non singular, with no ramification divisor.

A standard example of a bad orbifold is $\mathbf{P}^1$, with the ramification divisors 0 and ∞ and relatively prime ramification indices. It is its own universal covering.

14.12. Remark If $\mu = (\mu_1, \ldots, \mu_N)$ is a ball N-uple, i.e. $0 < \mu_i < 1$ for all i and $\Sigma_i \mu_i = 2$, and if in addition μ satisfies condition

INT: *for all distinct* i, j *with* $1 - \mu_i - \mu_j > 0$, $(1 - \mu_i - \mu_j)^{-1} \in \mathbf{Z}$,

then the space Q_μ^{st} can be regarded as an orbifold with ramification weight $1 - \mu_i - \mu_j$ attached to the divisor D_{ij} defined by the diagonal $x_i = x_j$ in $(P^1)^N$. The result of [DM] §10 can be interpreted as saying that Q_μ^{st} is an orbifold quotient of the complex ball if μ is a ball N-uple satisfying condition INT.

§15. ELLIPTIC AND EUCLIDEAN μ'S, REVISITED

15.1. Let $(\mu_s)_{s \in S}$ be a family of positive real numbers with sum 2. We continue to assume that $\#S = N = n + 3 \geq 4$. In [DM] §13, we defined μ to be elliptic (resp. euclidean) if for some element of S, which we will label ∞, one has $\mu_\infty > 1$ (resp. $\mu_\infty = 1$). With a slight change of the notation in [DM] §13.1, let M be the moduli space of injective maps $y : S \to \mathbf{P}^1$, with $y(\infty) = \infty$, taken modulo translations. Set $S_1 = S - \{\infty\}$. In [DM], instead of dividing by translations, we fixed $a \in S_1$ and imposed $y(a) = 0$. This would not be convenient here, as we will have to keep in mind the natural action of $\Sigma(S_1)$ on M. One can instead impose the condition that the barycenter of the $y(s)(s \in S_1)$ be zero, and unless otherwise stated, we shall realize M in that way:

$$M = \{\text{injective } y : S_1 \to \mathbf{C} \mid \sum_{s \in S_1} y(s) = 0\}.$$

We complete M as follows:

$$M_{st} = \{\text{all } y : S_1 \to \mathbf{C} \text{ with barycenter } 0\}.$$

We further introduce

$$Q_{st} = (M_{st} - \{0\})/\mathbf{G}_m,$$

the multiplicative group acting by multiplication. Q_{st} is a projective space of dimension n.

In [DM], we showed that when condition INT is satisfied, one has:
elliptic case: hypergeometric functions identify a finite ramified covering of M_{st} with $\mathbf{C}^{n+1}$, and a finite ramified covering of Q_{st} with a projective space.

euclidean case: hypergeometric functions identify a ramified covering of Q_{st} with $\mathbf{C}^n$.

Let us fix a subgroup H of the symmetric group $\Sigma(S_1)$. Assume that μ is H-invariant and that

(ΣINT) For s, t in S_1, either $1 - \mu_s - \mu_t$ is a reciprocal integer, or the transposition (s, t) is in H (hence $\mu_s = \mu_t$) and $\frac{1}{2}(1 - \mu_s - \mu_t)$ is a reciprocal integer.

The methods of [M3] allow us to generalize the results of [DM] §14 to that case, with M_{st} or Q_{st} replaced by their quotient by H. The proofs remain the same as in the loc. cit. and we will merely describe the results.

We first consider the elliptic case and, as an application show that the ball quotients constructed by Barthel-Hirzebruch-Hofer ([BHH], 1987) from the Hesse and extended Hesse line arrangements on $\mathbf{P}^2$ are quotients of the ball by groups commensurable with the monodromy groups Γ_μ for suitable ball 5-uples μ. An equivalent result was previously obtained by M. Yoshida ([Y], Theorem 2). We then consider the euclidean case, and as an application show that the ball quotients constructed by Barthel-Hirzebruch-Hofer (ibid) from line arrangements on abelian surfaces are quotients of the ball by groups commensurable with monodromy groups for suitable ball 5-uples μ.

Elliptic Case

15.2. The quotient $M_{st}/\Sigma(S_1)$ is the moduli space of divisors of degree $n + 2$ and barycenter 0 on the affine line A. Let $D \subset (M_{st}/\Sigma(S_1)) \times A$ be the universal divisor and let $f : P_{M_{st}/\Sigma(S_1)} \to M_{st}/\Sigma(S_1)$ be the fibering of its complement. It is a family of punctured projective lines: the fiber above the image $y \in M_{st}/\Sigma(S_1)$ of $\tilde{y} \in M_{st}$ is the complement in $\mathbf{P}^1$ of the point at infinity and of the $\tilde{y}(s)$ $(s \in S_1)$. In coordinates: $M_{st}/\Sigma(S_1)$ is the affine space with the elementary symmetric functions $\sigma_2, \cdots, \sigma_{n+2}$ of the $\tilde{y}(s)$ $(s \in S_1)$ as coordinates, and if t is the standard coordinate on A, the equation of D is

$$t^{n+2} + \sigma_2 t^n - \sigma_3 t^{n-1} \cdots + (-1)^{n+2}\sigma_{n+2} = 0.$$

For X mapping to $M_{st}/\Sigma(S_1)$, we denote by f_X (or simply f): $P_X \to X$ the pull back to X of $f : P_{M_{st}/\Sigma(S_1)} \to M_{st}/\Sigma(S_1)$.

Assume that μ is elliptic and that $H \subset \Sigma(S_1)$ fixes μ. The map $f_{M/H}$: $P_{M/H} \to M/H$ is a topological fibration. On $P_{M/H}$, we consider the rank one local system L trivialized on $(M/H) \times \mathbf{R}^+$ near ∞ and with fiberwise monodromy $\alpha_s = exp(2\pi i\mu_s)$ around $\tilde{y}(s)$, in the fiber P_y above the image y in M/H of $\tilde{y} \in M$.

For e the trivializing section of L on $\mathbf{R}^+$ near ∞,

$$w = \prod_{s \neq e}(z - \tilde{y}(s))^{-\mu_s} dz \; e$$

defines a cohomology class in $H^1(P_y, L)$. For variable y, one obtains a holomorphic section w_μ of the local system $R^1 f_{M/H*}L$.

Fix $0 \in M/H$. Let $(M/H)^{\sim}$ denote the covering of M/H corresponding to the monodromy of $R^1 f_{H*}L$. The section w_μ gives a map

$$\tilde{w}_\mu : (M/H)^{\sim} \to H^1(P_0, L)$$

homogeneous of degree $1 - \sum_{s \in S_1} \mu_s$ relative to the action of the multiplicative group $y \mapsto \lambda y, M \to M$. Let $(M_{st}/H)^{\sim}$ be the completion of the spread $\rho : (M/H)^{\sim} \to M/H$ over M_{st}/H.

THEOREM 15.3. *Assume that μ is elliptic, H-invariant, and satisfies ΣINT. Then $(M/H)^{\sim}$ is a finite covering of M/H and $\tilde{w}_\mu$ extends to an isomorphism from $(M_{st}/H)^{\sim}$ to $H^1(P_0, L_0)$.*

If $\Gamma'_{\mu,H}$ is the monodromy group of $R^1 f_{M/H*}L$, we conclude that the inverse $\tilde{w}_\mu^{-1}$ of $\tilde{w}_\mu$ induces an isomorphism

$$(15.3.1) \qquad H^1(P_0, L_0)/\Gamma'_{\mu,H} \xrightarrow{\sim} M_{st}/H.$$

The map $\rho\tilde{w}_\mu^{-1}$ ramifies only above the divisors $y(s) = y(t)$. The local monodromy of the local system $R^1 f_{M/H*}L$ around this divisor is a complex reflection, with non-trivial eigenvalue $exp((2\pi i)(\mu_s + \mu_t))$ if the transposition (s,t) is not in H, and $-exp(2\pi i\mu_s)$ otherwise. The ramification index of $\rho\tilde{w}_\mu^{-1}$ above $y(s) = y(t)$ is

$$(1 - \mu_s - \mu_t)^{-1} \text{ if the transposition } (s,t) \text{ is not in } H$$
$$2(1 - \mu_s - \mu_t)^{-1} \text{ if the transposition } (s,t) \text{ is in } H.$$

The condition (ΣINT) ensures that those numbers are integers.

Define $D = (1 - \sum_{s \in S_1} \mu_s)^{-1}$.

COROLLARY 15.4. *(i)* $D = (1 - \sum\limits_{s \in S_1} \mu_s)^{-1}$ *is an integer*

(ii) The dilations in $\Gamma'_{\mu,H}$ *form the group of* D^{th} *roots of unity.*

(iii) The order of $\Gamma'_{\mu,H}$ *is* $\#H \cdot D^{n+1}$.

(iv) If H is a product of symmetric groups acting on the parts of a partition of S_1, *i.e. if H is generated by transpositions, then* $\Gamma'_{\mu,H}$ *is generated by complex reflections. The reflection in* $\Gamma'_{\mu,H}$ *is of order* $(1 - \mu_s - \mu_t)^{-1}$ *for the transposition* $(s,t) \notin H$ *and* $2(1 - \mu_s - \mu_t)^{-1}$ *for* $(s,t) \in H$.

PROOF. For λ in a contractible neighborhood U of $1 \in \mathbf{C}^*$, the automorphism of M_{st}/H induced via multiplication by λ on M_{st} lifts to $(M_{st}/H)^\sim$, the lifting being unique if one requires continuous dependence on λ, and that the lifting be taken to be the identity for $\lambda = 1$. For λ in U and $\lambda^{1/D}$ the determination which is 1 for $\lambda = 1$, one has then a commutative diagram

$$
\begin{array}{ccc}
(M_{st}/H)^\sim & \longrightarrow & H^1(P_0, L) \\
\downarrow{\scriptstyle\lambda} & & \downarrow{\scriptstyle\lambda^{1/D}} \\
(M_{st}/H)^\sim & \longrightarrow & H^1(P_0, L).
\end{array}
$$

Reversing the horizontal isomorphism and replacing λ by $\lambda^{1/D}$, one gets

$$
\begin{array}{ccc}
H^1(P_0, L) & \longrightarrow & M_{st}/H \\
\downarrow{\scriptstyle\lambda} & & \downarrow{\scriptstyle\lambda^D} \\
H^1(P_0, L) & \longrightarrow & M_{st}/H
\end{array}
$$

By analytic continuation, this holds for all $\lambda \in \mathbf{C}^*$ and any determination λ^D. Using that $n \geq 1$, one checks that H does not contain any dilation. The automorphism λ^D of M_{st}/H hence uniquely determines the number λ^D and $\lambda \mapsto \lambda^D$ is a function: D is an integer.

The points x and λx will have the same image in M_{st}/H for all $x \in H^1(P_0, L)$ if and only if $\lambda^D = 1$. This proves (ii).

To prove (iii), we may replace H by a bigger group respecting μ. Taking for H the stabilizer of μ, we may and shall assume that H is as in (iv). In this case, M_{st}/H is non singular. By homogeneity, the map

$$
\rho\, \tilde{w}_\mu^{-1} : H^1(P_0, L) \to M_{st}/H
$$

maps 0 to 0, and it is totally ramified there, i.e., 0 is fixed by $\Gamma'_{\mu,H}$. The non singularity of M_{st}/H hence implies that $\Gamma'_{\mu,H}$ is generated by complex reflections, of orders the ramification indices above the divisors $y(s) = y(t)$ (the only divisors above which $\tilde{w}_\mu$ is not etale). This proves (iv).

Let T be the partition of S_1 such that H is the product of the $\Sigma(T)$ for $T \in \mathcal{T}$. The quotient M_{st}/H then admits as coordinates the elementary symmetric functions σ_i^T in the $y(t)$ $(t \in T)$, for $T \in \mathcal{T}$ - with the constraint $\sum_T \sigma_1^T = 0$. Instead of putting a constraint, one may suppress one σ_1^T.

The map $\rho\,\tilde{w}_\mu^{-1}$ is a finite map between non singular spaces and hence is flat. Its degree is the order of $\Gamma'_{\mu,H}$. The scheme theoretical inverse image of 0 is a non reduced finite scheme Y located at 0, whose length $\#Y$ is $\#\Gamma'_{\mu,H}$. It is defined by the $n+1$ equations $\sigma_i^T \circ \rho\tilde{w}_\mu^{-1}$ (one $\sigma_1^T \circ \rho\tilde{w}_\mu^{-1}$ omitted), homogeneous of degree $D \cdot i$. By Bezout, the length of Y is the product of the degrees:

$$\#\Gamma'_{\mu,H} = \#Y = \Pi \qquad D \cdot i = D^{n+1} \cdot \#H.$$

This proves (iii).

We call the fixed point set of a complex reflection its *mirror*.

15.5. Here is the list of pairs (μ, H) with $n \geq 2$ to which the theorem applies. We give the dimension n, the common denominator d of the μ_i, the integers $d\mu_i (i \in S_1)$, D (so that $\mu_\infty = 1 + \frac{1}{D}$), H, the numbering of $\Gamma'_{\mu,H}$ in Shephard-Todd [ST] and the Coxeter diagram [C]. Justification follows the list.

n	d	$d\mu_i$					D	H	$\#$	diagram
2	6	1	1	1	1		3	$\Sigma(4)$	#25	3 3 3
2	6	1	1	1	2		6	$\Sigma(3)$	#26	3 3 4
3	6	1	1	1	1	1	6	$\Sigma(5)$	#32	3 3 3 3

In M_{st}/H, let Y be the divisor where some $y(s)$ coincide. The map $H^1(P_0, L) \to M_{st}/H$ ramifies only above Y, with ramification index along the components of Y given by the μ's and, as $H^1(P_0, L)$ is simply connected, it is universal among such ramified coverings. In particular, $\Gamma'_{\mu,H}$ is a quotient of $\pi_1(M_{st}/H - Y)$: the quotient in which the local monodromy is of order given by the ramification index.

In the case labelled #25, $(M_{st}/H, Y)$ is deduced from a root system of type A_3 via dividing by the Weyl group, with Y the image of the mirrors. The fundamental group $\pi_1(M_{st}/H - Y)$ is the braid group on four strings, with each of the three standard generators conjugate to a local monodromy. This justifies the given diagram and locates $\Gamma'_{\mu,H}$ in Shephard-Todd list.

The case labelled #32 is similar, with five strings.

For the case labelled #26, A_3 is to be replaced by B_3; i.e., one has to consider the space of 4-uples (y_1, y_2, y_3, y_4), modulo translation, divided by $\Sigma(3)$. Instead of dividing by the translations, one fixes $y_4 = 0$. Put $y_i = z_i^2 (i = 1, 2, 3)$. Then, the z-space, divided by the Weyl group of type B_3 (permutations and sign changes) is the same as the y-space, divided by $\Sigma(3)$, and Y is the image of the mirrors: $z_s = \pm z_t$ and $z_s = 0$. The $\Gamma'_{\mu,H}$ is now a quotient of the generalized braid group of type B_3, justifying the given diagram.

The linear group $\Gamma'_{\mu,H}$ has 12 mirrors of order 3 in the cases #25 and #26; $\Gamma'_{\mu,H}$ has an additional 9 mirrors of order 2 in case #26.

15.6. Let us now divide by $\mathbf{G}_m$. One obtains that the projective space $\mathbf{P}H^1(P_0, L)$
$:= (H^1(P_0, L) - \{0\})/\mathbf{G}_m$ is a ramified covering of Q_{st}/H. The divisors along which it ramifies are the image of the $y_s = y_t$ as well as additional fixed point divisors of H acting on Q_{st}. There are such additional divisors only for $n = 2$, in which case $h = (ab)(cd)$ has the fixed point divisor $y_a = -y_b, y_c = -y_d$ in Q_{st}.

The cases #25 and #26 give rise to the same configuration of mirrors in $\mathbf{P}^2$ and hence the same configuration of divisors in the weighted projective plane $\mathbf{P}(1, 2, 3)$: the one considered in §11.6. The universal covering ramified with degree three along the cuspidal cubic $C : y^2 = x^3$ and degree two along the line L is a projective plane. The ramification divisor in $\mathbf{P}^2$ is the extended Hesse configuration of 21 lines: the projective mirrors corresponding to the 12 mirrors of order 3 of groups #25 or #26 form a Hesse configuration. The remaining 9 lines correspond to the mirrors of order 2 of the linear group #26. The mirrors of order 3 (resp. 2) are above C (resp. L).

An analogous configuration of divisors in weighted projective space $\mathbf{P}(2, 3, 4)$ was obtained by M. Yoshida by a direct computation, based on the knowledge of the polynomial invariants of the group #26.

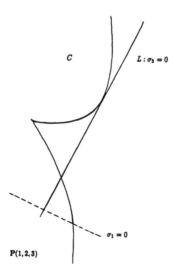

If we treat $\mathbf{P}^2$ as an orbifold, giving to the extended Hesse configuration mirrors of order 3 (resp 2) a weight α (resp. β), this orbifold is an orbifold-etale covering of $\mathbf{P}(1, 2, 3)$, with weight $\frac{\alpha}{3}$ (resp. $\frac{\beta}{2}$) attached to C (resp. L). These are the weights of C and L in the orbifold $Q_\mu^{st}/\Sigma(3)$ associated to the ball 5-uple μ with

$$\mu_1 = \mu_2 = \mu_3 = \frac{1}{2} - \frac{\alpha}{3}$$

$$\mu_3 + \mu_4 = 1 - \frac{\beta}{2}, i.e.$$

$$\mu = (\frac{1}{2} - \frac{\alpha}{3}, \frac{1}{2} - \frac{\alpha}{3}, \frac{1}{2} - \frac{\alpha}{3}, \frac{1}{2} + \frac{\alpha}{3} - \frac{\beta}{2}, \frac{2}{3}\alpha + \frac{\beta}{2})$$

As long as $\mu_1 + \mu_5 > 1$ and $\mu_4 + \mu_5 > 1$, $Q_\mu^{st}/\Sigma(3)$ is $\mathbf{P}(1, 2, 3)$. For $\mu_1 + \mu_5 < 1$ (resp. $\mu_4 + \mu_5 < 1$) one has to modify $\mathbf{P}(1, 2, 3)$ at the tangency point of C and L (resp. at the cusp of C). The modification is not a simple blowing up: the blowing up is done before dividing by $\Sigma(3)$.

In Barthel-Hirzebruch-Hofer, the following cases of ball quotients are considered:

in 5.7, $E : (\alpha, \beta) = (\frac{1}{2}, 1), (\frac{1}{3}, 1), (\frac{1}{4}, 1)$. Here $\beta = 1$: no ramification is allowed along mirrors of order 2; only order 3 mirrors, forming the Hesse configuration, are weighted.

In 5.7 $F : (\alpha, \beta) = (\frac{1}{2}, \frac{1}{2})(\frac{1}{2}, \frac{1}{3}), (\frac{1}{2}, 0), (\frac{1}{3}, \frac{1}{9}), (\frac{1}{4}, \frac{1}{2}), (\frac{1}{4}, \frac{1}{6}), (0, \frac{1}{3})$. As in loc.cit., we introduce m_4 and m_5 by

$$\frac{1}{m_4} = 1 - 2\alpha = 2(1 - \mu_4 - \mu_5)$$

$$\frac{1}{m_5} = \frac{1}{2}(3 - 3\beta - 2\alpha) = 3(1 - \mu_1 - \mu_5).$$

If $\frac{1}{m_4} > 0$ (resp. $\frac{1}{m_5} > 0$) the 4-fold points (resp. 5-fold points) of the configuration have to be blown up, the exceptional curve being weighted by $\frac{1}{m_4}$ (resp. $\frac{1}{m_5}$). If $\frac{1}{m_4}$ or $\frac{1}{m_5}$ is 0, the corresponding point has to be deleted. The 4-fold points are the intersections of 4 3-mirrors. The 5-fold points are the intersections of 2 3-mirrors and 3 2-mirrors, and one can have $\frac{1}{m_5} \geq 0$ only for $\beta \neq 1$. One can show that the 4-fold points of the extended Hesse configuration map to the cusp of C in $\mathbf{P}(1, 2, 3)$, that the 5-fold points map to the point of tangency of C and L, and that, in all cases, the ball quotient considered by Barthel-Hirzebruch-Hofer is an orbifold etale covering of $Q_\mu^{st}/\Sigma(3)$ (cf. §14). What is to be checked is that the modification of $\mathbf{P}(1, 2, 3)$ obtained by blowing up the 4-fold points of the extended Hesse configuration, then dividing by the group # 25 (projectively the same as # 26), is the same as the modification obtained by blowing up the point of Q_μ^{st} corresponding to $x_1 = x_2 = x_3$, then dividing by $\Sigma(3)$ — and similarly for 5-fold points and $x_1 = x_2 = x_4$ (and its $\Sigma(3)$ transforms). One should further check that the weights on these new curves match.

Euclidean case

15.7. Let H be a subgroup of $\Sigma(S_1)$ fixing μ. We assume μ euclidean: $\mu_\infty = 1$. Let $\bar{P}_{M/\Sigma(S_1)}$ be the complement in $M/\Sigma(S_1) \times \mathbf{P}^1$ of the universal divisor $D \subset M/\Sigma(S_1) \times A^1$. The complement in $\bar{P}_{M/\Sigma(S_1)}$ of the ∞-section is $P_{M/\Sigma(S_1)}$. We write $\bar{P}_X$ for a pull-back to X over $M/\Sigma(S_1)$.

As $\mu_\infty = 1$, the local system L on $P_{M/H}$ has trivial monodromy around ∞. It can be described as the restriction to $P_{M/H}$ of the local system, still noted L, on $\bar{P}_{M/H}$, trivialized on the ∞-section with fiberwise monodromy $exp(2\pi i \mu_s)$ around $\tilde{y}(s)$, in the fiber P_y above the image y in M/H of $\tilde{y} \in M$.

The multiplicative group $\mathbf{G}_m$ acts on M by $\lambda : y \mapsto \lambda y$ and on $P_M :$ $P_y \to P_{\lambda y}$ is multiplication by λ. The action descends to M/H and $P_{M/H}$

and the local system L on $P_{M/H}$ is equivariant.

Let M' be the open subset of M where the action of $H \times \mathbf{G}_m$ is free; i.e. $(M/H)' := M'/H$ is the open subset of M/H where $\mathbf{G}_m$ acts freely and $Q' := M'/\mathbf{G}_m$ is the open subset of Q where H acts freely. Dividing by $\mathbf{G}_m$, one sees that P_M, L and $R^1 f_* L$ descend to Q'/H. The holomorphic section w_μ of $R^1 f_* L$ defined by the form w of (15.2) is invariant by $M \times \mathbf{G}_m$ and descends as well.

The divisors in $M_{st} - M'$ are as follows: (cf. (8.3.2))

a) The divisor D_{ab} where $y(a) = y(b)$ (a, b distinct in S_1)

b) For $n = 2$ and $(a,b)(c,d) \in H$, the divisor where $y(a) = -y(b)$ (and hence $y(c) = -y(d)$).

Fix $0 \in (Q'/H)$. Let $(Q'/H)^\sim$ denote the covering of Q'/H corresponding to the monodromy of $(R^1 f_* L)_{Q'/H}$. The section w_μ gives a map

$$\tilde{w}_\mu : (Q'/H)^\sim \to H^1(P_0, L).$$

The form w_μ on P_y has residue -1 at infinity. If D is a small disc around ∞ and $D^* = D - \{\infty\}$, the cohomology class of w_μ induces on D^* $(-2\pi i)$ times the standard generator of $H^1(D^*, L) = H^1(D^*, \mathbf{C})$. It follows that $\tilde{w}_\mu$ maps $(Q'/H)^\sim$ to the affine hyperplane $Res_\infty = -1$ in $H^1(P_0, L)$:

$$\tilde{w}_\mu : (Q'/H)^\sim \to H^1(P_0, L)^{Res=-1}.$$

This map is etale.

Let $(Q_{st}/H)^\sim$ be the completion of $(Q'/H)^\sim$ over Q_{st}/H.

PROPOSITION 15.7. *If the condition ΣINT is satisfied, the map $\tilde{w}_\mu$ extends to an isomorphism*

$$(15.7.1) \qquad\qquad (Q_{st}/H)^\sim \xrightarrow{\sim} H^1(P_0, L)^{Res\ =-1}$$

The affine space $H^1(P_0, L)^{Res\ =-1}$ is parallel to the homogeneous vector subspace $H^1(P_0, L)^{Res\ =0}$. The latter is simply $H^1(\bar{P}_0, L)$, $\bar{P}_0 := \mathbf{P}^1 - 0(S_1)$, a cohomology group on which the intersection form is definite.

The monodromy group Γ acts on $H^1(P_0, L)^{Res\ =-1}$ by affine transformations, respecting the unitary structure of $H^1(\bar{P}_0, L)$. Inverting (15.7.1) gives an isomorphism

$$(15.7.2) \qquad\qquad \tilde{w}_\mu^{-1} : H^1(P_0, L)^{Res\ =-1}/\Gamma \xrightarrow{\sim} Q_{st}/H.$$

The quotient is compact and, by Bieberbach's theorem, Γ has a subgroup of finite index of translations. For a modern account of Bieberbach's theorem, see [L. Auslander 1960]. Let Γ_0 be the subgroup of translations in Γ. The quotient
$H^1(P_0, L)^{Res\ =-1}/\Gamma_0$ is a complex torus, and a finite covering of Q_{st}/H. It is hence an abelian variety. This is an abuse of language for "homogeneous space over an abelian variety": there is no natural origin.

LEMMA 15.8. *If, for some partition T of S_1, H is the corresponding product of symmetric groups $\Sigma(T)(T \in \mathcal{T})$, then the complement U in Q_{st}/H of any complex codimension 2 subvariety is simply connected.*

PROOF. The inverse image U_1 of U in the projective space Q_{st} is the complement of a codimension two subspace. The simple connectedness of the projective space Q_{st} hence implies that of U_1. Let $\tilde{U}$ be a connected covering of U. Its pull back $\tilde{U}_1$ to U_1 is a covering of U_1. It hence has a section s:

$$
\begin{array}{ccc}
\tilde{U}_1 & \longrightarrow & \tilde{U} \\
\downarrow{\scriptstyle s} & & \downarrow \\
U_1 & \longrightarrow & U
\end{array}
$$

and U_1 dominates $\tilde{U}$: $\tilde{U} = U_1/G$ for some subgroup G of H. It follows that $U_1/G \to U$ is etale, hence that G contains any $h \in H$ having a fixed point set of codimension 1. As H is generated by such h (the transpositions in H), $G = H$, and $U = U_1/H = U_1/G = \tilde{U}$; i.e., any connected covering of U is trivial and U is simply connected.

Remark For $n = 2$, the lemma can be strengthened in either of the following two ways

a) Remove from U the trace of a divisor $y_a = -y_b, y_c = -y_d$.

b) Take H generated by transpositions and elements of the Vierergruppe.

PROPOSITION 15.9. *For H as in the lemma, the monodromy group $\Gamma :=$ $\Gamma_{\mu,H}$ acting on $H^1(P_0, L)^{Res\ =-1}$ is generated by reflections. Consequently the group of linear parts, isomorphic to Γ/Γ_0, is a finite group generated by complex reflections.*

PROOF. The lemma ensures that Γ is generated by local monodromy transformations around the divisors of Q_{st}/H in the complement of Q'/H;

they are complex reflections on

$$H^1(P_0, L)^{Res=-1}.$$

Remark 15.10 a) around the divisor $y(s) = y(t)$, the monodromy is a reflection of order the denominator of $1 - \mu_s - \mu_t$ if $(s, t) \notin H$, and the denominator of $\frac{1}{2}(1 - \mu_s - \mu_t)$ if $(s, t) \in H$.

b) For $n = 2$, and $(ab)(cd) \in H$ the monodromy around the divisor $y_a = -y_b, y_c = -y_d$ is of order 2.

15.11 Assume condition ΣINT, define $\Delta = \Gamma/\Gamma_0$, and let A be the abelian variety $H^1(P_0, L)^{Res=-1}/\Gamma_0$. By (15.7.2)

$$(15.11.1) \qquad\qquad A/\Delta \xrightarrow{\sim} Q_{st}/H$$

Fix $a \in A$, with image $\bar{a}$ in Q_{st}/H. Let $\Delta_a \subset \Delta$ be the fixer of a. The map (15.11.1) induces a local isomorphism, at a, from A/Δ_a to Q_{st}/H. It follows that Δ_a is the local fundamental group, at $\bar{a}$, of the orbifold Q_{st}/H; the divisors in the complement of Q'/H have to be weighted according to 15.10. If Q_{st}/H is non-singular at $\bar{a}$, Δ_a is generated by complex reflections. If further Δ_a is maximal among finite linear groups on $\mathbb{C}^{dim\ A}$ generated by complex reflections, it follows from 15.9 that $\Delta = \Delta_a$. In that case, Γ is the semi-direct product of Δ_a and Γ_0.

15.12. Assume $n \geq 2$. Fix $a, b \in S_1$ and let $\bar{S}$ be deduced from S by identifying a and b; one has $\pi : S \to \bar{S}$. We define $\bar{\mu}$ on $\bar{S}$ by $\bar{\mu}(x) = \sum_{\pi(y)=x} \mu(y)$. The divisor $D_{ab} : y(a) = y(b)$ in Q_{st} is the space $Q_{st}(\bar{S})$: Q_{st} for $\bar{S}$ and the euclidean $\bar{\mu}$. Let $\bar{H}$ be the image in $\Sigma(\bar{S}_1)$ of the subgroup of H stabilizing $\{a, b\}$. The quotient $Q_{st}(\bar{S})/\bar{H}$ maps to the image D_{ab}/H of D_{ab} in Q_{st}/H; as $Q_{st}(\bar{S})/\bar{H}$ is normal and the map is finite and birational, $Q_{st}(\bar{S})/\bar{H}$ is the normalization of $D_{a,b}/H$.

Along $D_{a,b}$, the local monodromy of the local system $R^1 f_* L$ is a complex reflection. The fixed part extends over D_{ab}, where it is the local system corresponding to $\bar{\mu}$ on $Q(\bar{S})/\bar{H}$. The section w_μ tends to $w_{\bar{\mu}}$.

Let $D'_{a,b}/\bar{H}$ be a Zariski dense smooth open subset of the image of D_{ab} in Q_{st}/H. Let $(D'_{a,b}/\bar{H})^{\sim}$ be a connected component of its inverse image in $(Q_{st}/H)^{\sim}$ and let $(Q_{st}(\bar{S})/\bar{H})^{\wedge}$ be its completion over $Q_{st}(\bar{S})/\bar{H}$. For

M the fixed mirror, in $H^1(P_0, L)^{Res} = -1$, of local monodromy around $D_{a,b}$, we get a commutative diagram

(15.12.1)

$$
\begin{array}{ccc}
(Q_{st}(\bar{S})/\bar{H})^\wedge & \xrightarrow{\ w\ } & M \\
\downarrow{\scriptstyle i} & & \downarrow \\
(Q_{st}/H)^\sim & \xrightarrow{\ w_\mu\ } & H^1(P_0, L)^{Res} = -1
\end{array}
$$

We now assume ΣINT for μ. The map w_μ is then an isomorphism. As i is generically injective (it is on $(D'_{ab}/\bar{H})$) w is generically injective. The map w is a determination of $w_{\bar{\mu}}$ and it follows that $\bar{\mu}$ satisfies ΣINT and that the covering $(D'_{a,b}/\bar{H})^\sim$ is given by the monodromy of the local system corresponding to $\bar{\mu}$. The diagram can be rewritten

(15.12.2)

$$
\begin{array}{ccc}
(Q_{st}(\bar{S})/\bar{H})^\sim & \xrightarrow{\ w_{\bar{\mu}}\ } & M \\
\downarrow{\scriptstyle i} & & \downarrow \\
(Q_{st}/H)^\sim & \longrightarrow & H^1(P_0, L)^{Res} = -1
\end{array}
$$

and induces isomorphisms

(15.12.3)

$$
\begin{array}{ccc}
Q_{st}(\bar{S})/\bar{H} & \xrightarrow{\ \sim\ } & M/\Gamma_{\bar{\mu},\bar{H}} \\
\downarrow & & \downarrow \\
Q_{st}/H & \xrightarrow{\ \sim\ } & H^1(P_0, L)/\Gamma_{\mu,H}
\end{array}
$$

The left vertical map is generically injective. It follows that $\Gamma_{\bar{\mu},\bar{H}}$ is the image of the stabilizer of M :

$$
\Gamma_{\bar{\mu},\bar{H}} = \text{stabilizer } (M)/\text{fixer } (M) \text{ in } \Gamma_{\mu,H}.
$$

In particular,

(a) $\Gamma_0(\bar{\mu})$ consist of those translations in $\Gamma_{\mu,H}$ stabilizing M

(b) The stabilizer of M induces on M a group generated by complex reflections.

It follows from (a) that the abelian variety $B := M/\Gamma_0(\bar{\mu})$ is a subabelian variety of $A := H^1(P_0, L)/\Gamma_0$. In terms of $\Delta(\bar{\mu}) = \Gamma_{\bar{\mu},\bar{H}}/\Gamma_0$, (15.12.3) can be rewritten

$$
\begin{array}{ccccc}
Q_{st}(\bar{S})/\bar{H} & \xrightarrow{\ \sim\ } & B/\Delta(\bar{\mu}) & \longleftarrow & B \\
\downarrow & & \downarrow & & \downarrow \\
Q_{st}/H & \xrightarrow{\ \sim\ } & A/\Delta & \longleftarrow & A
\end{array}
$$

with $\Delta(\bar{\mu})$ the stabilizer in Δ of B, modulo the fixer of B. From this diagram we can deduce

LEMMA 15.13. *Suppose $y \in Q_{st}(\bar{S})/\bar{H}$ has an image x in Q_{st}/H which is the image of the unique point $a \in A$ fixed by Δ and $A \to Q_{st}/H$ totally ramified above x. Then*

(i) y is the only point in $Q_{st}(\bar{S})/\bar{H}$ mapping to x

(ii) $a \in B$ and, in B, is fixed by $\Delta(\bar{\mu})$; $B \to Q_{st}(\bar{S})/\bar{H}$ is totally ramified above y.

PROOF.

$$\{a\} \cap B = \bigcup_{y' \to x} (\text{inverse image of } y' \text{ in } B)$$

15.14. We will apply 15.11, 15.12, 15.13 to relate the Example 1.4A of Barthel-Hirzebruch-Hofer 1987 to hypergeometric functions. The euclidean 7-uple

$$\mu = (\frac{1}{6} \ \frac{1}{6} \ \frac{1}{6} \ \frac{1}{6} \ \frac{1}{6} \ \frac{1}{6} \ 1)$$

satisfies ΣINT for $H = \Sigma(6)$. In $Q_{st}/\Sigma(6)$, let x be the point corresponding to 6-uples $(x_1, \cdots, x_6)$ with $x_1 = x_2 = x_3 = x_4 = x_5$. Near x, $Q_{st}/\Sigma(6)$ is isomorphic to $Q_{st}/\Sigma(5)$: 6-uples of points, not all equal, modulo translations, dilations and $\Sigma(5)$. Instead of dividing by translation, one can impose $\sum_1^5 x_i = 0$. Instead of dividing by dilations, one imposes $x_6 = 1$. One obtains that near x, $Q_{st}/\Sigma(6)$ with the orbifold structure determined by μ, is isomorphic to $((\text{hyperplane } \Sigma x_i = 0 \text{ in } \mathbf{C}^5)/\Sigma(5), 0)$ with the orbifold structure obtained by giving the weight $\frac{1}{3}$ to the discriminant locus. This model corresponds to the elliptic $\nu := (\frac{1}{6}\frac{1}{6}\frac{1}{6}\frac{1}{6}\frac{1}{6}, \frac{7}{6})$. It then follows that the local fundamental group at x is the complex reflection group $\Gamma'_{\nu, \Sigma(5)}$ of order $5!6^4$, and Coxeter diagram $\overset{\bullet \ \bullet \ \bullet \ \bullet}{3 \ 3 \ 3 \ 3}$ (#32 in Shephard Todd, see 15.5). Comparing its order with that of the other primitive reflection groups, one checks it is maximal. It follows that

$$Q_{st}/\Sigma(6) = A/\Delta$$

for Δ acting on an abelian variety A with a fixed point 0, corresponding to x in $Q_{st}/\Sigma(6)$, and acting on the tangent space at 0 as the complex reflection group #32.

The discriminant divisor in $Q_{st}/\Sigma(6)$ is irreducible. It follows that all mirrors M for the action of $\Gamma_{\mu,H}$ on affine space are conjugate.

We now apply 15.12, 15,13 to the divisor $x_5 = x_6$. One has

$$\bar{\mu} = (\frac{1}{6}, \frac{1}{6}, \frac{1}{6}, \frac{1}{6}, \frac{1}{3}, 1)$$
$$\bar{H} = \Sigma(4), \bar{S}_1 = \{1, 2, 3, 4, 5\}$$

and the point x in $Q_{st}/\Sigma(6)$ is the image of the point y of $Q_{st}(\bar{S})\Sigma(4)$ corresponding to $x_2 = x_3 = x_4 = x_5$. We have

$$Q_{st}(\bar{\mu})/\Sigma(4) = B/\Delta(\bar{\mu}),$$

where $\Delta(\bar{\mu})$ has a fixed point in B, corresponding to y.

Let us apply 15.12, 15.13 again, to the divisor $x_4 = x_5$. We obtain for $\bar{\bar{S}}_1 = \{1, 2, 3, 4\}$.

LEMMA 15.15. *For the euclidean* $\mu := (\frac{1}{6}, \frac{1}{6}, \frac{1}{6}, \frac{1}{2}, 1)$ *and the point* x *corresponding to* $x_2 = x_3 = x_4$, *one has*

$$Q_{st}/\Sigma(3) = A/\Delta,$$

and Δ *has a fixed point* 0 *in* A, *corresponding to* x *in* $Q_{st}/\Sigma(3)$.

We will use the point 0 as origin in A to define the group law of A.

15.16 Fix μ as in 15.15. The space $Q_{st}/\Sigma(3)$ is the weighted projective space $\mathbf{P}(1, 2, 3)$ of §11.6 with the cuspidal cubic C corresponding to $x_1 = x_2$, and the line L, which is tangent to C, corresponding to $x_3 = x_4$. For the orbifold structure corresponding to μ, both C and L have weight $1/3$. At x, we have simple tangency between C and L, and the orbifold structure near x is the same as that given by the elliptic ΣINT - satisfying $\nu := (\frac{1}{6}, \frac{1}{6}, \frac{1}{2}, \frac{7}{6})$ for $H = \Sigma(2)$ i.e. $\Delta \approx \Gamma'_{\nu, \Sigma(2)}$. It follows from 15.4 that Δ is of order $2.6^2 = 72$ and that the dilations in Δ are the 6th roots of 1. The group Δ, acting on the tangent space of A at 0, is an extension of the tetrahedral group in $PGL(2) = SO(3)$ (order 12) by μ_6.

PROPOSITION 15.17. *There is a isomorphism between the abelian variety* A *of 15.15 and* $(\mathbf{C}/\mathcal{O})^2, \mathcal{O} = \mathbf{Z}[\sqrt[3]{1}]$, *such that*

(i) The inverse image of $L \subset Q_{st}/\Sigma(3)$ *is the union of four elliptic curves through* 0. *The slopes of their tangent space are*

$$\lambda = 0, 1, exp(2\pi i/6) \text{ and } \infty.$$

(ii) Δ is the group of automorphisms of A, stabilizing the system of elliptic curves (i).

(iii) The inverse image of $C \subset Q_{st}/\Sigma(3)$ is the union of the four elliptic curves $y = \lambda x, \lambda = \frac{1}{2} + \frac{\sqrt{-3}}{3}, \frac{1}{2} - \frac{\sqrt{-3}}{3}, 1 + \frac{\sqrt{-3}}{3}$ and $\frac{\sqrt{-3}}{3}$.

(iv) Let U denote the subgroup of elements of order dividing 2 in any one of the elliptic curves lying over C. Then the intersection of any two of the elliptic curves above C is U. Moreover, the inverse image of the cusp point of C is $U - \{0\}$.

PROOF. We have $A = T/H$, for T the tangent space at 0 and H a lattice in T. This lattice is stable by Δ, hence by μ_6: it is a $\mathcal{O}$-module in T. The class number of $\mathcal{O}$ being 1, it is a free $\mathcal{O}$-module. Choosing a basis of H over $\mathcal{O}$, we obtain an isomorphism between A and $(\mathbf{C}/\mathcal{O})^2$.

LEMMA 15.18. *The inverse image in A of L is the union of four elliptic curves in A, any two of them meeting only in 0.*

Assuming 15.18, we now prove 15.17 (i). Let E_1, E_2, E_3, E_4 be the four elliptic curves above L. As E_1 and E_2 meet only in 0, the addition map $E_1 \times E_2 \to A$ is an isomorphism. As E_3 meets E_1 and E_2 only in 0, it is a graph of an isomorphism between E_1 and E_2. If we choose an isomorphism of E_2 with $\mathbf{C}/\mathcal{O}$, we get that $A = (\mathbf{C}/\mathcal{O})^2$, in such a way that the Lie algebra of $E_i (i = 1, 2, 3)$ is the line in $\mathbf{C}^2$ of slope $0, \infty$ and 1 respectively. The elliptic curve E_4, similarly, is the graph of an isomorphism $E_1 \to E_2$: it has an equation $y = \lambda x$ for λ a unit in $\mathcal{O}$, i.e. a 6th root of 1. Further, E_3 and E_4 meet only in 0. This means that $\lambda - 1$ is a unit. The only possibilities are $\lambda = exp(\pm 2\pi i/6)$. Exchanging E_1 and E_2, we may assume that the sign is +.

An automorphism of A preserving $\{E_1, E_2, E_3, E_4\}$ induces an even permutation of them which preserve the cross ratio of the tangent lines at 0. If the induced permutation is the identity, the action on the tangent space at 0 must be multiplication by a scalar. Preserving the lattice H, it must be a sixth root of 1. The order of the automorphism group is hence at most 12 times 6. As this is the order of Δ, Δ is the full automorphism group of $(A, \{E_1, E_2, E_3, E_4\})$.

The inverse image of C is again a union of images of mirrors. As $A \to Q_{st}/H$ totally ramifies at 0, each irreducible component of the inverse image goes through 0. The mirrors are of order 3. In A, they must correspond to elliptic curves E, and the tangent space at 0 must have a slope which is a fixed point of an element of order 3 of the alternating group A_4. The group A_4 is the group of projectivities of $\mathbf{P}^1$ stabilizing $\{0, 1, \infty, \frac{1+\sqrt{-3}}{3}\}$ and, computing the fixed points of elements of order 3, one gets (iii).

PROOF OF 15.18. By 15.12, the inverse image of L in A is a union of elliptic curves. It suffices to show that this inverse image is singular only above x, i.e. at 0. This is a local assertion on Q_{st}/H. Only above singular points of $C \cup L$ or of Q_{st}/H could the inverse image be singular. Besides x, there are two such points: (1) the transversal intersection point of C and L, and (2) the point at infinity of L, defined by $\sigma_1 = \sigma_3 = 0$ in the notation of §11, (whose inverse image in A consists of 2-division points). We consider them in turn.

(15.18.1) In suitable local coordinates u, v, the equation of L (resp C) is $u = 0$ (resp. $v = 0$). The local universal covering of the orbifold is obtained by extracting the third roots of u and v: it is the (U, V)-plane, with $(U, V) \mapsto (u, v) = (U^3, V^3)$. The inverse image of L is given by the equation $U = 0$, hence is non singular.

(15.18.2) The space Q_{st}/H has here a quadratic cone singularity, with L a non singular curve through the vertex. A quadratic cone, near the vertex, is the quotient of the (u, v)-plane by the involution $(u, v) \mapsto (-u, -v)$, and the inverse image L' of L is a non singular curve stable by that involution. To get the local universal covering, one further extracts a cube root of an equation for L'. The pull back of L' is non singular.

Let E denote an elliptic curve in A through 0. Let $U_2(A)$ and $U_2(E)$ denote the set of points in A and E respectively of order exactly 2. Then $U_2(A)$ has 15 elements and the homothety group μ_6 permutes transitively the 3 points of $U_2(E)$. If E is one of the elliptic curves above L, then by Lemma 15.18 the Δ orbit of $U_2(E)$ consists of 12 points; let U' denote the complement of these 12 points in $U_2(A)$. If E' is one of the elliptic curves above C, then $U_2(E') \subset U'$ since the points of order 2 in lines above L lie above the point at infinity of L, which is not in $C \cap L$. Hence $U - \{0\} = U_2(E') = U'$ for any of the four elliptic curves over C. The inverse image of C can have a singularity above only the cusp or the point of tangency of C and L (for by the argument of 15.18.1 applied to C, one can ignore the transversal intersection point of C and L); i.e. any two elliptic curves over C can intersect in points above the cusp of C or at 0. Inasmuch as U' is a Δ orbit, it is the complete inverse image of the cusp point of C, and the intersection of any two elliptic curves above C is precisely U. This proves (iv).

15.19 As in 15.6, if we now turn A into an orbifold by assigning weight α (resp. β) to the elliptic curves above L (resp. C), we obtain an orbifold-etale covering of the orbifold Q_{st}/H, with weight $\frac{\alpha}{3}$ and $\frac{\beta}{3}$ attached to L and C. Such weights correspond to a ball 5-uple μ with

$$\mu_1 = \mu_2 = \mu_3 = \frac{1}{2} - \frac{\beta}{3}$$

$$\mu_3 + \mu_4 = 1 - \frac{\alpha}{3}$$

i.e.

$$\mu = (\frac{1}{2} - \frac{\beta}{3}, \frac{1}{2} - \frac{\beta}{3}, \frac{1}{2} - \frac{\beta}{3}, \frac{1}{2} + \frac{\beta}{3} - \frac{\alpha}{3}, \frac{2\beta}{3} + \frac{\alpha}{3}).$$

The ball quotient case considered by Barthel-Hirzebruch-Hofer is $(\alpha, \beta) = (1/3, 1)$, corresponding to $(\frac{3}{18}, \frac{3}{18}, \frac{3}{18}, \frac{13}{18}, \frac{14}{18})$. They further blow up the origin and assign to it the weight $\frac{1}{m_4} = 1 - 2\alpha = \frac{1}{3}$. One has $\mu_1 + \mu_5 < 1$, so that Q_μ^{st}/H for μ is a blow up of the Q_{st}/H considered so far in our euclidean case. With the orbifold structure defined by the ball 5-uple μ, it should admit as orbifold-etale covering the orbifold considered by Barthel-Hirzebruch-Hofer; to see what should be checked, cf. the end of 15.6.

15.20. The other abelian surface considered by Barthel-Hirzebruch-Hofer 1.4, similarly corresponds to the euclidean μ

$$\mu = (\frac{1}{4}, \frac{1}{4}, \frac{1}{4}, \frac{1}{4}, 1),$$

with $H = \Sigma(3)$. The orbifold to be considered is as before, except that C has now weight $\frac{1}{4}$ and L weight $\frac{1}{2}$. The abelian variety A of which Q_{st}/H is a quotient A/Δ is now as follows.

PROPOSITION 15.21. *(i) Δ acting on A has a fixed point 0, which we take as origin. It maps to the cuspidal point of C*

(ii) A is isomorphic to $(\mathbf{C}/\mathcal{O})^2$ with $\mathcal{O} = \mathbf{Z}[i]$, and the isomorphism can be taken such that (iii)-(v) below holds.

(iii) The inverse image of C is the union of six elliptic curves through 0. The slopes λ of their tangent space are $\lambda = 0, 1 + i, 1, i, \frac{1+i}{2}, \infty$.

(iv) The inverse image of L is the union of the six elliptic curves which are the translates of a curve E as in (iii) by a $(1 + i)$-division point not lying on it.

(v) Δ is the group of automorphisms of A preserving the system of 6 lines (iii).

SKETCH OF PROOF. At the cusp x of C, the orbifold Q_{st}/H is locally isomorphic to the one corresponding to the elliptic ν $(\frac{1}{4}, \frac{1}{4}, \frac{1}{4}, \frac{5}{4})$, $H = \Sigma(3)$. By 15.4, the local fundamental group, which acts as $\Gamma'_{\nu, \Sigma(3)}$, is of order $6.4^4 = 96$, an extension of the octahedral group in $PGL(2) = SO(3)$ by μ_4. This group is maximal among groups generated by reflection of order 2 and 4. (i) follows.

Since the group Δ contains μ_4, acting on the tangent space at 0 by multiplication and since $\mathbf{Z}[i]$ has class number one, A is isomorphic to $(\mathbf{C}/\mathbf{Z}[i])^2$.

The inverse image of C is singular only at 0 and above the tangency point y of C and L. At y, the orbifold is isomorphic to the one given by the elliptic ν $(\frac{1}{4}, \frac{1}{4}, \frac{1}{4}, \frac{5}{4})$, $H = \Sigma(2)$: the local fundamental group has order $2 \cdot 4^2 = 32$, hence the fiber above y consist of $3 = 96/32$ points. At each, the inverse image of C consist of two smooth curves meeting transversally. It follows that the six elliptic curves through 0 occur in three pairs, with two of them meeting at a point other than 0 if and only if they belong to the same pair. The proof of (iii) is now parallel to that of 15.16 (i), and that of (v) to that of 15.16 (ii). The group Δ acts on the six elliptic curves (ii) as the symmetry group of the cube acts on the six faces of the cube.

The curves above L are fixed by some element of Δ. They hence are translates of the fixed curve through 0: the curves (iii). They meet curves (iii) above y, i.e. at the intersection of two of them. Those intersection points are the $(1 + i)$-division points. This proves (iv).

15.22 As in 15.19, we should now consider orbifold structures on A with weight α and β attached to the elliptic curves over C (resp. L). On Q_{st}/H, this implies weight $\frac{\alpha}{4}$ for C and $\frac{\beta}{2}$ for L, corresponding to the ball 5-uple

$$\mu = (\frac{1}{2} - \frac{\alpha}{4}, \frac{1}{2} - \frac{\alpha}{4}, \frac{1}{2} - \frac{\alpha}{4}, \frac{1}{2} + \frac{\alpha}{4} - \frac{\beta}{2}, \frac{\alpha}{2} + \frac{\beta}{2}).$$

The ball quotient case considered by Barthel-Hirzebruch-Hofer is $(\alpha, \beta) =$ $(\frac{1}{2}, 1)$: $\mu = (\frac{3}{8}, \frac{3}{8}, \frac{3}{8}, \frac{1}{8}, \frac{6}{8})$. The origin in A has to be blown up, with weight $\frac{1}{2}$ attached to the exceptional curve, and the orbifold so obtained should be (cf. end of 15.6) an orbifold-etale covering of the orbifold $Q_\mu^{st}/\Sigma(3)$ of §14.

Remark 15.23 In the setting described in 15.16 and 15.17, one can give a simple definition of the degree 4 map $P(1,2,3) \rightarrow P(1,2,3)$ which is defined in §§11.2-11.15 from first principles.

Let U denote the intersection of the four elliptic curves in A above C, and let Δ denote the automorphism group of A stabilizing the set of elliptic curves over L. Then Δ stabilizes U. Set $A' = A/U$ and let Δ' denote the automorphism group of A' induced by Δ. The properties of the group Δ used in the proof of Proposition 15.17 apply to Δ', thus yielding isomorphisms

$$A \simeq (\mathbf{C}/\mathbf{Z}[\sqrt[3]{1}])^2 \simeq A', \ A/\Delta \simeq A'/\Delta'$$

in which the elliptic curves in A over C (resp. L) map to the elliptic curves in A' over L (resp. C). The composition

$$\mathbf{P}(1,2,3) = Q_{st}/\Sigma(3) \simeq A/\Delta \rightarrow (A/U)/\Delta' \simeq A/\Delta \simeq \mathbf{P}(1,2,3)$$

is the degree 4 map described in §11.2.

§16. LIVNE'S CONSTRUCTION OF LATTICES IN $PU(1,2)$

Let $H := \{\tau \in \mathbf{C}; Im \ \tau > 0\}$ be the upper half plane and let $\Gamma(n) :=$ $\{\gamma \in SL(2,\mathbf{Z}); \gamma \equiv 1(mod \ n)\}$ denote the principal congruence subgroup of $SL(2,\mathbf{Z})$ of level n, $n \geq 3$. The semi-direct product $SL(2,\mathbf{Z}) \times (\frac{1}{n}\mathbf{Z})^2$ operates discontinuously on $H \times \mathbf{C}$ via

$$\left(\begin{pmatrix} a & b \\ c & d \end{pmatrix}, (x,y) : (\tau,z) \rightarrow (\frac{a\tau+b}{c\tau+d}, \frac{z+x\tau+y}{c\tau+d}) \right)$$

Set

$$E(n)^0 = (\Gamma(n) \times \mathbf{Z}^2) \backslash H \times \mathbf{C}, \ X(n)^0 = \Gamma(n) \backslash H.$$

The map $p : E(n)^0 \rightarrow X(n)^0$ is a fibering with elliptic curves as fibers, and points in the base space are in bijective correspondence with isomorphism classes of elliptic curves E together with an isomorphism $\alpha : (\mathbf{Z}/n\mathbf{Z})^2 \xrightarrow{\sim} E_n$, the group of division points of order n, such that α can be lifted to an isomorphism $\alpha : \mathbf{Z}^2 \rightarrow H_1(E,\mathbf{Z})$ respecting the orientation of $H_1(E,\mathbf{Z})$; such data define a *special level n structure* on E. K. Kodaira has studied the completion of an elliptic fibration to a relatively minimal non-singular algebraic surface: no component of its fibers can be blown down yielding a non-singular surface. Livne's construction begins with the relatively minimal completion $E(n)$ of $E(n)^0$, which in the terminology of Deligne-Rapaport is a *generalized elliptic curve* on the base $X(n) := \Gamma(n) \backslash (H \cup P^1(\mathbf{Q}))$, the complete curve consisting of $X(n)^0$ and $\frac{1}{2} n^2 \prod_{p|n}(1 - p^{-2})$ cusps. That is,

$$p : E(n) \rightarrow X(n)$$

is a morphism for which every fiber is either an elliptic curve or a Neron n-gon (over the field $\mathbf{C}$, this is Kodaira's special fiber $_1I_n$) := an n-cycle of projective lines $\Theta_0 + \Theta_1 + \ldots + \Theta_{n-1}$ with $(\Theta_i, \Theta_i) = -2, (\Theta_i, \Theta_{i+1}) = 1, i = 0, 1, \ldots n - 1$ (cf. [K], p.123)

We denote by D_i, $i \in (\mathbf{Z}/n\mathbf{Z})^2$, the section corresponding to an n-division point.

In [L], Livne proves the following. Fix $n \geq 3$. Set $d = n/gcd(n,6)$.

0.

$$D_i^2 = \frac{-n^3}{24} \prod_{p|n}(1 - p^2)$$

(loc. cit. §1.2 (7) pg. 6)

1. There is a cyclic covering $\tilde{E}_d(n)$ of $E(n)$ with group μ_d and standard d-uple ramification along the D_i and nowhere else, satisfying

(a) The action of each element of the group $\bar{A} := SL(2, \mathbf{Z}/n\mathbf{Z})$ $\times (\frac{1}{n}\mathbf{Z}/\mathbf{Z})^2$ on $E(n)$ lifts to $\tilde{E}_d(n)$ (loc. cit. Theorem 4).

(b) Let A denote $Aut \; \tilde{E}_d(n)$, the automorphism group of $\tilde{E}_d(n)$. Then (loc. cit. Theorem 10) A is a central extension:

$$1 \to \mu_d \to A \to SL(2, \mathbf{Z}/n\mathbf{Z}) \times (\mathbf{Z}/n\mathbf{Z})^2 \to 1;$$

the extension of $(\mathbf{Z}/n\mathbf{Z})^2$ by μ_d is given by the Weil pairing.

2. For $n = 7, 8, 9$, or 12, $\tilde{E}_d(n)$ is a compact quotient of the ball by a torsion-free subgroup of its automorphisms.

3. For $n = 5$, the lift of each D_i has self-intersection -1; when we contract these, the resulting surface is a compact quotient of the ball.

In the case of both (2) and (3), the proof is by verifying the $c_1^2 = 3c_2$ criterion for a 2-ball quotient. Livne denotes by $F_{n,d}$ the group of all lifts of elements of $Aut \; \tilde{E}_d(n)$ to the complex ball i.e.

$$1 \to \pi_1(\tilde{E}_d(n)) \to F_{n,d} \to Aut \; \tilde{E}_d(n) \to 1.$$

As in Livne, $\tilde{D}_i$ denotes the curve in $\tilde{E}_d(n)$ above the n-section D_i, $i \in (\mathbf{Z}/n\mathbf{Z})^2$.

Remark There is a clerical error dropping an exponent 2 on page 102 (Theorem 7, assertion 3) of Livne's thesis which survives in the presentation on page 108 for $F_{n,d}$; relation R_5) should read "$y^{4d} = 1$" not "$y^{2d} = 1$".

Next we relate Livne's groups $F_{n,d}$ to the monodromy groups $\Gamma_{\mu,\Sigma}$ for 5-uples

$$\mu = (\frac{1}{2} - \frac{1}{n}, \frac{1}{2} - \frac{1}{n}, \frac{1}{2} - \frac{1}{n}, \frac{1}{n}, \frac{1}{2} + \frac{2}{n}).$$

For $n > 4$, one has $0 < \mu_i < 1$ for $1 \leq i \leq 5$. For $4 \geq n > 2$, one has $0 < \mu_i < 1$ for $i < 5$ and $\mu_5 \geq 1$, with $\mu_5 = 1$ for $n = 4$. For $n > 4$, the number $1 - \mu_i - \mu_j$ and its sign is as follows:

$i,j \in \{1,2,4\}$:	$\frac{2}{n}$	positive
$i \in \{1,2,3\}, j=4$:	$\frac{1}{2}$	positive
$i \in \{1,2,3\}, j=5$:	$-\frac{1}{n}$	negative
$(i,j)=(4,5)$:	$\frac{n-6}{2n}$	$n>6$: positive; $n=6$: zero; $n<6$, negative

This data tells the coincidences allowed for the moduli space Q_μ^{sst} of 5-uples of points on the projective line corresponding to the 5-uple μ. We set $\Sigma =$ the permutation group of $\{1,2,3\}$. When Q_μ^{st}/Σ will be considered as an orbifold, the weight of the divisor where $x_i = x_j$ is taken to be $1 - \mu_i - \mu_j$, divided by 2 if $i,j \in \{1,2,3\}$ (cf. §14).

Let $E(n)_{rem}$ denote the space obtained by removing the divisors D_i from $E(n)$, and let $E(n)_{con}$ denote the compactification of $E(n)_{rem}$ obtained by contracting each D_i to a point. We define the spaces $\tilde{E}(n)_{rem}$ and $\tilde{E}(n)_{con}$ similarly, starting from $\tilde{E}_d(n)$ and the $\tilde{D}_i$.

THEOREM 16.1. *With the notations as above,*

(i) μ satisfies condition ΣINT for $n = 5,6,7,8,9,10,12,18$, and for these n,

$$Ball/\Gamma_{\mu\Sigma} = Q_\mu^{st}/\Sigma$$

as orbifolds.

(ii) $F_{n,d}$ is conjugate in $PU(1,2)$ to $\Gamma_{\mu\Sigma}$, for $n = 7,8,9,12$.

(iii) For $n = 7,8,9,12$, $\tilde{E}_d(n)/A = Q_\mu^{st}/\Sigma$ as orbifolds.

(iv) For $n = 5$, $\tilde{E}(5)_{con}$ is a ball quotient, and $\tilde{E}(5)_{con}/A = Q_\mu^{st}/\Sigma$ as orbifolds.

(v) For $n = 6$, the ball covers $E(6)_{rem}$, $E(6)_{rem}/\bar{A} = Q_\mu^{st}/\Sigma$ as orbifolds, and $E(6)_{con}/\bar{A} = Q_\mu^{sst}/\Sigma$.

(vi) For $n = 10,18$, $\tilde{E}_d(n)/A = Q_\mu^{st}/\Sigma$, and the ball is a ramified covering of $E(n)/(\mathbf{Z}/n\mathbf{Z})^2 \cdot (\pm 1)$ but not of $E(n)$.

(vii) For $n = 4$, $\tilde{E}(4)_{con}$ is an abelian surface, and $\tilde{E}(4)_{con}/A = Q_\mu^{st}/\Sigma$ as orbifolds (cf. 15.7.2).

(viii) For $n = 3$, we have $d = 1$, $A = \bar{A}$, $E(3)_{con}$ is the projective plane, and $E(3)_{con}/A = Q_\mu^{st}/\Sigma$ as orbifolds (cf. 15.6).

PROOF. (i) is a special case of the main result in [M3] restated for orbifolds (cf. §14). The proof of the remaining assertions consists of comparing modular interpretations of $\tilde{E}_d(n)/A$ and of Q_μ^{sst}/Σ. We will then see that these are isomorphic orbifolds when $n = 7, 8, 9, 12$. In the remaining cases $n = 5, 6, 10, 18$ for which μ satisfies the ΣINT condition, Livne's $\tilde{E}_d(n)/A$ must be reinterpreted in order to arrive at an orbifold isomorphic to Q_μ^{st}/Σ, i.e. to obtain an orbifold whose orbifold-universal covering is the ball.

Livne's group $F_{n,d}$ can be described as the orbifold-fundamental group of the orbifold quotient $\tilde{E}_d(n)/A$; to each image of a divisor $\tilde{D}$ in $\tilde{E}_d(n)$ whose fixer $\{g \in A; gx = x$ for all $x \in \tilde{D}\}$ is non-trivial, there is assigned the ramification index equal to the order of the fixer.

We compute in stages the quotient $\tilde{E}_d(n)/A$ and the ramification indices assigned to its ramification divisors.

1. $\tilde{E}_d(n)/\mu_d = E(n)$. The ramification divisors are the n-division point sections, each with ramification index d.

2. $E(n)/(\mathbf{Z}/n\mathbf{Z})^2$

   (a) on any fiber E above a point of $X(n)^0$ we have

$$1 \rightarrow E_n \rightarrow E \xrightarrow{n} E \rightarrow 1$$

Hence $E(n)^0/(\mathbf{Z}/n\mathbf{Z})^2 \xrightarrow[z \rightarrow nz]{\sim} E(n)^0$.

   (b) Above each cusp of $X(n)$, one gets the 1-gon:

a rational curve intersecting itself. In a neighborhood of infinity, the equation of the fiber can be taken as $y^2 = 4x^3 - g_2 x - g_3$ with discriminant $g_2^3 - 27g_3^2 = 0$ but $(g_2, g_3) \neq (0,0)$ at the cusp or equivalently $j := \frac{g_2^3}{g_2^3 - 27g_3^2} = \infty$ at the cusp. The involution $z \rightarrow -z$ of each fiber is given by $(x, y) \rightarrow (x, -y)$.

Set $E(n)' = E(n)/(\mathbf{Z}/n\mathbf{Z})^2$. This orbifold is a fibering over $X(n)$ and has only one ramification divisor: the 0-section with ramification d.

3. Division by $(\pm 1) \subset SL(2, \mathbf{Z}/n\mathbf{Z})$ yields $E(n)'/(\pm 1)$. For each elliptic fiber E of $E(n)'$, $E/(\pm 1)$ is a projective line on which are marked

   0—the image of the 0-section

   $\mathcal{A}$—the image of the points of order 2, a three element subset.

Thus we get in $E(n)'/(\pm 1)$ two branch divisors denoted

   $0 :=$ the image of the 0-section of $E(n)'$

   $\mathcal{A} :=$ the irreducible divisor which meets each fiber over

   a point of $X(n)^0$ in the image of points of order 2.

The three elements in which $\mathcal{A}$ meets a generic fiber cannot be marked if n is odd because they are permuted by monodromy. The ramification index along 0 is $2d$ and along $\mathcal{A}$ is 2. At the special fibers over the points of $X(n) - X(n)^0$, the discriminant is zero and two of the three branches of the divisor $\mathcal{A}$ come together.

To assert that $p : E(n) \to X(n)$ is a generalized elliptic curve is to say that there is a morphism

$$+ : E(n)^{reg} \times_{X(n)} E(n) \to E(n)$$

where $E(n)^{reg} := \{u \in E(n); p^{-1}(p(u))$ is non-singular $u\}$ such that
   (a) the restriction of $+$ to $E(n)^{reg}$ yields a commutative group scheme
   (b) the morphism $+$ is the group action of $E(n)^{reg}$ on $E(n)$
   (c) For every cusp $c \in X(n)$, the translation $v \to u + v$ of $u \in p^{-1}(c)^{reg}$ on $p^{-1}(c)$ induces a rotation on the graph of the special fiber $p^{-1}(c)$.

Livne, following Mumford, calls $E(n)$ the universal elliptic curve of level n.

Indeed, Deligne-Rapoport show that $p : E(n) \to X(n)$ is the universal generalized elliptic curve with special structure of level n specified by an isomorphism

$$\alpha : (\mathbf{Z}/n\mathbf{Z})^2 \xrightarrow{\sim} E_n^{reg}$$

liftable to $\mathbf{Z}^2 \to H_1(E_1\mathbf{Z})$ where E_n^{reg} denotes the group scheme of the n-division points.

The action of $SL(2, \mathbf{Z}) \times (\frac{1}{n} \mathbf{Z}^2)$ on $H \times \mathbf{C}$ descends to an action of $\bar{A} \cong SL(2, \mathbf{Z}/n) \times (\mathbf{Z}/n\mathbf{Z})^2$ on $E(n)^0$ and this action extends to $E(n)$ by the unicity of the relatively minimal model $E(n)$.

The action of $(\mathbf{Z}/n)^2$ on $E(n)$ is given by the prescribed isomorphism α.

Each element of E_n^{reg} yields a section of $E(n)$ over $X(n)$ and each component of a special fiber meets n of these n^2 sections in n distinct points.

4. The last division is by $PSL(2, \mathbf{Z}/n\mathbf{Z})$. Here the base $X(n)$ is divided by $PSL(2, \mathbf{Z}/n\mathbf{Z})$ giving as quotient the j-line. We can regard the j-line as the moduli space of a projective line P with four points:

A marked point 0 and an unordered set $\mathcal{A}$ of three points A', A'', A'''; any two elements of $\mathcal{A}$ can coincide but 0 cannot coincide with an element of $\mathcal{A}$. Namely, the set of isomorphism classes of such $(P; 0; A', A'', A''')$ is in bijective correspondence with the set of isomorphism classes of the elliptic curve which is a double covering of the projective line ramified at 0 and at the elements of $\mathcal{A}$. Thus $(E(n)'/(\pm 1))/PSL(2, \mathbf{Z}/n\mathbf{Z})$ is the fiber space above the j-line with fiber at $(P; 0; A', A'', A'')$ equal to $P/Aut(P; 0; A', A'', A'')$. In other words, the quotient $\tilde{E}_d(n)/A$ is the moduli space of a projective line, a point 0, an unordered set of three points A', A'', A''' and an additional point x. The allowed coincidences are: two elements of $\mathcal{A}$, or $x = 0$, or $x =$ one or two elements of $\mathcal{A}$. The fiber map $\tilde{E}_d(n)/A \to X(n)/A$ is given by the projection forgetting x.

The ramification divisors with their ramification are:

The 0-section: $x = 0$, ramification $2d$.

Points of order 2: $x = A', A''$, or A''', ramification 2

Fiber at $j = \infty$: Two elements of $\mathcal{A}$ coincide; ramification n.

This last ramification results from the fact that $\begin{pmatrix} 1 & 1 \\ 0 & 1 \end{pmatrix}$ fixes the cusp at $j = \infty$ and each point of the special fiber above it. These are the only branch divisors.

On the other hand, the moduli space of five points A', A'', A''', $x, 0$ on a projective line with $\{A', A'', A'''\}$ unordered, taken with the foregoing data for coincident points is Q_μ^{sst}/Σ where $\mu = (\frac{1}{2} - \frac{1}{n}, \frac{1}{2} - \frac{1}{n}, \frac{1}{2} - \frac{1}{n}, \frac{1}{n}, \frac{1}{n}, \frac{1}{2} + \frac{2}{n})$, $\Sigma = \Sigma(\{A', A'', A'''\})$, and $n > 6$. Set $\mu_A = \frac{1}{2} - \frac{1}{n}, \mu_x = \frac{1}{n}, \mu_0 = \frac{1}{2} + \frac{2}{n}$. For $n > 2$, condition ΣINT holds for μ if and only if $n \in \mathbf{Z}$ and $\mu_x + \mu_0 = 1 - \frac{1}{\tau}$ with $\tau \in \mathbf{Z}$; that is, $\tau = \frac{2n}{n-6} \in \mathbf{Z}$. If $n > 4$ and μ

satisfies condition ΣINT, then by (i), $\Gamma_{\mu\Sigma}$ is the orbifold-universal covering group of the orbifold-ball quotient Q_μ^{st}/Σ.

From the foregoing computation of $\tilde{E}_d(n)/A$, one can extract the following:

$n > 6$: $\tilde{E}_d(n)/A = Q_\mu^{st}/\Sigma$

$n = 6$: $E(6)_{rem}/A = Q_\mu^{st}/\Sigma$ (here $d = 1, E(6) = \tilde{E}_d(6), A = \bar{A}$), and the space $E(6)_{rem}$ is a ball quotient; taking the one point compactification of both sides, one gets

$$E(6)_{con}/A = Q_\mu^{sst}/\Sigma$$

$n = 5$: $\widehat{E(5)}_{con}/A = Q_\mu^{st}/\Sigma$.

For $n = 5, 6, 7, 8, 9, 12$, these isomorphisms with Q_μ^{st}/Σ are isomorphisms of orbifolds. That the weights agree on the divisors where $x = A', A''$ or A''' or where two A''s coincide is clear (see the list of $1 - \mu_i - \mu_j$ preceding Theorem 16.1): for $n = 5, 6$ there are no other divisors to consider; for $n = 7, 8, 9, 12$, one observes that $gcd(6, n) = n - 6$ so the Livne's $d :=$ $n/gcd(6, n) = n/n - 6$ and thus $d = \tau/2$, a positive integer since τ is even in these four cases. This proves (ii), (iii), (iv), (v).

Remark 16.2 Given n, Livne's choice of d was made as a consequence of three results:

1. $d \mid (n/gcd(2, n))$ assures the existence of cyclic covering with Galois group μ_d ramified along the D_i and nowhere else (loc.cit §1.4, Theorem 1, pg. 16).

2. $d \mid (n/gcd(6, n))$ assures that each element of $Aut\ E(n)$ lifts to $Aut\ \tilde{E}_d(n)$. (loc. cit. §2.3, Theorem 4, pg. 41).

3. $c_1^2/c_2 = 3$ if and only if $(d - 1)\frac{n}{d} = 6$ (loc. cit. §1.6, Proposition, pg. 22). If we solve this linear equation for d, we find

$$d = \frac{n}{n - 6} = \tau/2.$$

In order to distinguish this choice of d from Livne's $d = \frac{n}{gcd(6,n)}$, we set $d' = \tau/2$.

Remark 16.3 One can formally consider orbifolds with ramification indices which are arbitrary real numbers, interpreting the negative and infinite indices of divisors as indicating they must be contracted to a point.

In that generality, all the assertions (iii) through (vii) of Theorem 16.1 express the same theme:

If we regard $E(n)$ as an orbifold with weight $1/d'$ along each of the D_i, then the orbifold quotient $E(n)/\bar{A}$, suitably modified at divisors with weights ≤ 0, is isomorphic to the orbifold Q_μ^{st}. More explicitly, for $n = 5, d' = -5$, and by the formula of Livne cited above, $D_i^2 = -5$. Hence in $\tilde{E}_5(5)$, $\widetilde{D_i^2} = \frac{1}{5}D_i^2 = -1$ for each $i \in (\mathbf{Z}/n\mathbf{Z})^2$. Blowing down each $\tilde{D}_i$ then dividing by the group to which A descends yields the orbifold isomorphism $\tilde{E}(5)_{con}/A = Q_\mu^{st}/\Sigma$ of (iv). Similar remarks hold for $n = 3, 4, 6$. For $n = 6, 1/d' = 0$, and the divisors D_i must be removed. For $n = 4, D_i^2 = -2$ so that $\tilde{D}_i^2 = -1$. Hence one expects $\tilde{E}(4)_{con}/A = Q_\mu^{st}/\Sigma$. For $n = 3, D_i^2 = -1$. Accordingly one expects (viii).

When $n = 10$ of 18, τ is odd and we have to interpret $E(n)$ as a formal orbifold with ramification $\tau/2$ along each divisor D_i. One can replace $E(n)$ by $E(n)'/(\pm 1) := E(n)/(\pm 1) \times (\mathbf{Z}/n\mathbf{Z})^2$. Treating the latter as an orbifold, it has ramification index $2 \cdot \tau/2 = \tau$ along the image of the D_i and ramification index 2 along the image of the 2-division points in $E(n)/(\mathbf{Z}/n)^2$. Thus the ball is an orbifold covering of $E(n)'/(\pm 1)$ but the ball is not an orbifold covering space of $E(n)$.

Taken all together, we see that the orbifold quotient $E(n)/\bar{A}$ can be described in terms of the ball quotient orbifold Q_μ^{st}/Σ for all n such that μ satisfies condition ΣINT, except for $n = 3, 4$.

Remark 16.4 For $n = 3$, $E(3)_{con} \cong \mathbf{P}^2$ is the orbifold universal covering of Q_μ^{st}/Σ. The elliptic μ is the one we met in 15.6 with $\alpha = \beta = 1$. The group $\bar{A}$ of automorphisms of $E(3)_{con}$ is generated by complex reflections with the mirrors forming the extended Hesse configuration.

Remark 16.5 For $n = 4$, $\tilde{E}(4)_{con}/A = Q_\mu^{st}/\Sigma$ with μ the euclidean 5-uple of 15.20. Hence the abelian surface $\tilde{E}(4)_{con}$ is an orbifold covering of the Q_μ^{st}/Σ considered there.

§17. LINE ARRANGEMENTS: QUESTIONS

17.1. We first review some computations in Barthel, Hirzebruch and Höfer (1987). If the universal covering of a compact surface S is the complex ball $B : S = B/\Gamma$ with Γ acting freely, then

$$(17.1.1) \qquad\qquad \mathrm{prop}(S) := 3\chi(S) - K^2$$

vanishes. If a smooth irreducible divisor D of $S = B/\Gamma$ is covered by a subball of B, then

$$(17.1.2) \qquad\qquad \mathrm{prop}(D) := 2D^2 - \chi(D)$$

vanishes.

The same vanishings hold, more generally, if S admits a hermitian metric of constant holomorphic curvature and if D is a totally geodesic smooth divisor. Such a surface S, if not a ball quotient, is either a quotient of an abelian surface by a finite group acting freely, or the projective plane. In the first case, D is the image of an elliptic curve; in the second, D is a line.

17.2. If S is a compact quotient of the ball $S = B/\Gamma$, where the discrete group Γ acts properly, but not freely, similar identities hold in a suitable orbifold sense. Once orbifold analogues of prop have been defined, and the relation between its values for S and for a finite orbifold covering S_1 of S has been ascertained, the proof is by reduction to the free action case, using that if $\Gamma_1 \subset \Gamma$ is a subgroup of finite index acting freely on B, B/Γ_1 is a finite orbifold covering of B/Γ.

Let S be a compact orbifold with set of ramification divisors $\mathcal{D}$. Locally, it is the orbifold quotient U/Γ of a non singular space U by a finite group Γ. The *weight* of a point x is the inverse of the order of the stabilizer Γ_u in Γ of a point u of U above x. Example: the weight of any point of a ramification divisor D, neither on *Sing* S nor on *Sing* $\bigcup \mathcal{D}$, is the weight of D. Let T be a stratification of S such that on each stratum T the weight is constant, denoted $w(T)$. The strata T form a partition of S and one defines

$$(17.2.1) \qquad\qquad \chi_{orb}(S) := \Sigma w(T)\chi(T).$$

The Euler-Poincaré characteristic of an algebraic variety V, or more generally of a compact analytic space minus a subspace, is the same for cohomology and for cohomology with compact support: $\chi(V) = \chi_c(V)$. Hence χ is additive in V, because χ_c is: for a partition of V into strata W_j, one has $\chi_c(V) = \Sigma\chi_c(W_j)$. From this it follows that (17.2.1) is independent of the chosen stratification. If S_1 is a finite orbifold covering of S of degree d, one has

$$(17.2.2) \qquad\qquad \chi_{orb}(S_1) = d\ \chi_{orb}(S).$$

Let NS (S) be the Neron-Severi group of S. In NS $(S) \otimes \mathbf{R}$, one defines

$$(17.2.3) \qquad\qquad K_{orb} := K + \Sigma(1 - w(D))D$$

(sum over the ramification divisors). It makes sense: for N divisible enough, the line bundle $(\Omega^{dim\ S})^{\otimes N}(\Sigma N(1-w(D))D)$ on the complement of $Sing(S)$ extends, uniquely, to S. If $\pi : S_1 \to S$ is a finite orbifold covering of S, one has

$$(17.2.4) \qquad\qquad K_{orb}(S_1) = \pi^* K_{orb}(S).$$

Assume now that S is a compact surface. For $\pi : S_1 \to S$ a finite orbifold covering of degree d, it follows from (17.2.4) that

$$(17.2.5) \qquad\qquad K_{orb}(S_1)^2 = d\ K_{orb}(S)^2.$$

One defines as in (17.1.1)

$$(17.2.6) \qquad\qquad \mathrm{prop}_{orb}(S) = 3\chi_{orb}(S) - K_{orb}^2.$$

We keep assuming that S is a compact surface. Let D be a (Weil) divisor whose reduced inverse image on any orbifold chart is locally a union of smooth divisors, any two of which meet transversally. The following definition of $\text{prop}_{orb}(D)$ is concocted so that (a) for D the union of distinct irreducible divisors D_i, $\text{prop}_{orb}(D)$ is the sum of the $\text{prop}_{orb}(D_i)$; (b) for π : $S_1 \rightarrow S$ a finite orbifold covering of degree d, $\text{prop}_{orb}(\pi^* D) = d\,\text{prop}_{orb}(D)$. The definition is as in 17.1.2, with the following modifications. Let $\{D_i\}$ be the irreducible components of D, and $\{w_i\}$ their weights. For $x \in D$, choose an orbifold chart $\varphi : V \rightarrow S$ covering x and let $n(x)$ be the number of branches of $\varphi^* D$ meeting at a point above x. One decomposes D into open sets $U_i \subset D_i$ and a finite number of exceptional points, in such a way that the weight is constant - equal to w_i - on U_i and that U_i is orbifold-smooth, i.e. that $(\varphi^* U_i)_{red}$ is smooth in any orbifold chart φ. In (17.1.2), one replaces $\chi(D)$ by

$$(17.2.7) \qquad\qquad \Sigma w_i \chi(U_i) + \Sigma w(x) n(x).$$

One defines

$$(17.2.8) \qquad\qquad D_{orb} := \Sigma w_i D_i$$

and D^2 is replaced in (17.1.2) by

$$(17.2.9) \qquad\qquad (D_{orb})^2 - \Sigma w(x)\frac{n(x)(n(x) - 1)}{2}.$$

Both (17.2.7) and (17.2.8) have the required behavior in an orbifold covering $S_1 \rightarrow S$.

17.3. We are interested in the case where S is a non singular surface and where the ramification divisors D_i are smooth and in general position. By "general position", we mean that no point is on three of the divisors and that they intersect transversally. If S is an orbifold ball quotient B/Γ, the inverse image of D_i in B is then a disjoint union of subballs, and if D_i and D_j meet at x, D_i and D_j are covered by subballs meeting orthogonally at a point above x.

As stratification of S, one can use the decomposition into the complement U of the D_i (weight 1), the $U_i := D_i - \bigcup_{j \neq i} D_j$ (weight w_i) and the intersection points of the divisors ($x \in D_i \cap D_j$ has weight $w_i w_j$). One has

$$\chi_{orb}(S) = \chi(U) + \Sigma w_i \chi(U_i) + \sum_{i<j} w_i w_j D_i D_j$$

$$= \chi(S) - \Sigma(1 - w_i)\chi(D_i) + \sum_{i<j}(1 - w_i)(1 - w_j) D_i D_j$$

Let D be an irreducible divisor which is orbifold smooth, i.e. whose reduced inverse image in orbifold charts is smooth. Concretely, it means that D is one of the D_i, or that it is smooth and in general position together with them. In (17.2.7), all $n(x)$ are one. The quantity (17.2.7) is called $\chi_{orb}(D)$. For D of weight w one has then

$$(17.3.2) \qquad \chi_{orb}(D) = w(\chi(D) - \sum_{D_i \neq D} (1 - w_i)DD_i).$$

Applying the adjunction formula, one gets

$$(17.3.3) \qquad \chi_{orb}(D) = -D_{orb}(K_{orb} + D_{orb}),$$

with $D_{orb} = wD$ and K_{orb} as in 17.2.3. It follows that

$$\text{prop}_{orb}(D) = 2D_{orb}^2 - \chi_{orb}(D)$$
$$(17.3.4) \qquad\qquad = D_{orb}(K_{orb} + 3D_{orb})$$

$$(17.3.5) \qquad \frac{1}{w} \text{prop}_{orb}(D) = D(K_{orb} + 3\,D_{orb})$$

17.4. Let S be any smooth compact surface. Let D_i be smooth irreducible divisors in general position on S. Let w_i be real numbers, the *weight* of the D_i. We keep defining $\text{prop}_{orb}(S)$ and $\text{prop}_{orb}(D)$ by the formula 17.2.3, 17.2.5 and those of 17.3. For $D = D_i$, of weight $w = w_i$, $\frac{1}{w}\text{prop}_{orb}(D)$ is defined by (17.3.5), even if $w = 0$. If we let the w_i vary, $\text{prop}_{orb}(S)$ is quadratic (non homogeneous) in the w_i. The computation of [BHH] p. 178 can be presented as follows

LEMMA 17.4.1. $\partial\, \text{prop}_{orb}(S)/\partial w_i = \frac{-1}{w_i} \text{prop}_{orb}(D_i)$.

PROOF. One has $\partial K_{orb}/\partial w_i = -D_i$, hence

$$\frac{\partial}{\partial w_i} K_{orb}^2 = -2K_{orb}D_i = \frac{-2}{w_i} K_{orb}D_{i_{orb}}.$$

By (17.3.1), (17.3.2), and (17.3.3),

$$\frac{\partial}{\partial w_i}\chi(S)_{orb} = \chi(D_i) - \sum_{j \neq i}(1 - w_j)(D_i D_j) = \frac{1}{w_i}\chi_{orb}(D_i)$$

$$= -\frac{1}{w_i}D_{i_{orb}}(K_{orb} + D_{i_{orb}})$$

The left hand side is hence

$$\frac{1}{w_i}(2\ K_{orb}D_{iorb} - 3D_{iorb}(K_{orb} + D_{[iorb}))$$

$$= \frac{-1}{w_i}D_{iorb}(K_{orb} + 3D_{iorb}) = -D_i(K_{orb} + 3\ D_{iorb}),$$

which one compares to (17.3.5).

COROLLARY 17.4.2. *If for weights w_i^0 one has $prop_{orb}(S) = 0$, and all $prop_{orb}(D_i) = 0$, then for any $w_i = w_i^0 + \Delta w_i$, one has*

$$prop_{orb}(S) = \frac{1}{2}\sum_i \Delta w_i \cdot \frac{-1}{w_i}\ prop_{orb}(D_i)$$

In particular, if the $1/w_i\ prop_{orb}(D_i)$ vanish, so does $prop_{orb}(S)$.

By 17.4.1, $prop_{orb}(S)$, viewed as a function f of the Δw_i, is homogeneous quadratic, and one applies Euler's formula $\Sigma x_i \partial_{x_i} f = 2f$.

Let S be a smooth compact surface, $\pi = \tilde{S} \to S$ be the blow up of S at a finite number of points $(P_i)_{1 \le i \le n}$ and let E_i be the exceptional curves.

PROPOSITION 17.5. *If each E_i is given the weight -1, one has*

(17.5.1) $prop(S) = prop_{orb}(\tilde{S})$ *and*

(17.5.2) $prop_{orb}(E_i) = 0$

If D' is the pure transform a smooth irreducible divisor of D (to which is assigned weight 1), one has

(17.5.3) $prop(D) = prop_{orb}(D')$

PROOF. One has $\chi(\tilde{S}) = \chi(S) + n$ and $K(\tilde{S}) + \Sigma E_i$, hence

$$\chi_{orb}(\tilde{S}) = \chi(\tilde{S}) - \Sigma(1 - (-1))\chi(E_i)$$
$$= \chi(S) + n - 4n = \chi(S) - 3n$$
$$K_{orb}(\tilde{S}) = K(\tilde{S}) + \Sigma(1 - (-1))E_i$$
$$= \pi^* K(S) + \Sigma E_i + \Sigma 2E_i = \pi^* K(S) + 3\Sigma E_i$$
$$K_{orb}^2(\tilde{S}) = K(S)^2 + 9\Sigma E_i^2 = K(S)^2 - 9n$$

and (17.5.1) follows. One has

$$\chi_{orb}(E_i) = -\chi(E_i) = -2$$
$$(E_{i\ orb})^2 = (-E_i)^2 = E_i^2 = -1$$

and (17.5.2) follows. If the sums are extended over the j for which P_j is on D, and there are m of them,

$$\chi_{orb}(D') = \chi(D) - \sum_j (1 - (-1)) = \chi(D) - 2m$$
$$D'^2 = (\pi^* D - \Sigma E_j)^2 = D^2 + \Sigma E_j^2 = D^2 - m$$

and (17.5.3) follows.

17.6. Example. Let $\mathcal{C}$ be a line arrangement in the projective plane $\mathbf{P}^2$. Let $\mathbf{P}^{\sim}$ be deduced from $\mathbf{P}^2$ by blowing up all intersection points of three or more lines in $\mathcal{C}$. Let $\mathcal{C}'$ consist of the pure transforms of the lines in $\mathcal{C}$ (with weight 1) and of the exceptional curves (with weight -1). Then,

$$\text{prop}_{orb}(\mathbf{P}^{\sim}) = 0 \text{ and}$$
$$\text{prop}_{orb}(D) = 0 \text{ for } D \text{ in } \mathcal{C}'.$$

The same goes for a configuration of abelian curves on an abelian surface.

17.7. LEMMA. *Let V be a 2-dimensional complex vector space, $\Gamma \subset GL(V)$ a finite group and $S := V/\Gamma$. We assume that*

(a) S is non singular, i.e. Γ is generated by complex reflections.

(b) The ramification divisors of the orbifold V/Γ are smooth and meet transversally.

(c) There are at least three of them.

Then, if $\tilde{V}$ (resp. $\tilde{S}$) is deduced from V (resp. S) by blowing up the origin 0, one has $\tilde{V}/\Gamma = \tilde{S}$.

PROOF. There is a non-singular minimal model S_1 for $\tilde{V}/\Gamma$ (cf. Barth, Peters, Van de Ven, "Compact Surfaces", Springer Verlag (1984), Ch. III, Theorem 6.2). Inasmuch as the composition $S_1 \to \tilde{V}/\Gamma \to S$ is a bimeromorphic map, the surface S_1 is obtained from S by an iteration of blow-ups (ibid Corollary 4.4). Since S_1 is minimal and the fiber over 0 of $\tilde{V}/\Gamma \to S$ is irreducible, we see that

(a) S_1 is obtained by blowing up in turn points $P = 0$ of V/Γ, $P_1, \ldots, P_n$ with P_{i+1} on the inverse image E_i of P_i, and

(b) $\tilde{V}/\Gamma$ is obtained by blowing down the E_i $(i \neq n)$.

Let $B_1, \ldots, B_k$ be the ramification divisors $(k \geq 3)$. After 0 is blown up, their pure transforms meet E_0 in at least three distinct points, at most one of which is P_1. If $n \neq 0$, it follows that the pure transforms of $B_1, \ldots, B_k$ on $\tilde{V}/\Gamma$ meet the exceptional curve in at most two points (all but at most one meet at the singular point produced by the contraction). On the other hand, the inverse images of $B_1, \ldots, B_k$ in V are union of lines: the mirrors of complex reflections. Those lines correspond one to one with the intersection of their pure transform on $\tilde{V}$ with the exceptional curve, and it follows that the intersection of the B_i with the exceptional curve are disjoint in $\tilde{V}/K$ - contradicting $k \geq 3$. As $k \geq 3$, one must have $n = 0$, which proves the lemma.

17.8. Remark. Condition (c) rules out the case where $\mathbf{C}^2/\Gamma$ is $\mathbf{C}^2$, with coordinates lines as ramifications divisors. This corresponds to Γ being a group of diagonal matrices $(x, y)(x \in \mu_n, y \in \mu_n)$. In this case, the conclusion of 17.7 holds if and only if $n = m$.

17.9. Let $\Gamma \subset Aut(B)$ be a discrete group acting properly on the 2-dimensional complex ball B, with a compact quotient S. Let $\mathcal{D}$ be the set of ramification divisors of the orbifold B/Γ. We assume that S is smooth and that at each singular point of $\bigcup \mathcal{D}$, the branches of $\bigcup \mathcal{D}$ are smooth and meet transversally. Let Σ be the set of points of S where at least three branches of $\bigcup \mathcal{D}$ meet, and let $B^\sim$ be deduced from the ball by blowing up the points in the finitely many orbits corresponding to points of S in Σ. It follows from 17.7 that $\tilde{S} := \tilde{B}/\Gamma$ coincide with the blow up of S at the points of Σ. As a consequence: (a) $\tilde{S} := \tilde{B}/\Gamma$ is non singular. (b) The union of the image in $\tilde{S}$ of the exceptional curves and of the points with non trivial stabilizer in Γ has as irreducible components smooth divisors D_i in general position. The weight of D_i is defined to be the weight it has in the orbifold $\tilde{B}/\Gamma$, with changed sign for the image of the exceptional divisors.

If $\Gamma_1 < \Gamma$ is an invariant subgroup of finite index in Γ acting freely on B, one obtains this system of weights for $\tilde{B}/\Gamma$ by (a) blowing up in B/Γ_1 the points above Σ, and giving the exceptional curves the weight -1, (b) dividing in an orbifold sense by the finite group Γ/Γ_1: the weight of the image of an irreducible divisor D is the weight of D in $(B/\Gamma_1)^\sim$ divided by the order of its fixer. Applying 17.5 and using the behavior of prop in coverings, one obtains

17.10 PROPOSITION. *With the assumptions and rotations of 17.9, one has*

$$\text{prop } \tilde{S} = 0 \text{ and}$$
$$\text{prop } D_i = 0$$

Remark 17.11. We could as well have blown up in B/Γ points of $\bigcup \mathcal{D}$ where two branches meet, provided the two branches have the same weight. This follows from 17.8.

17.12. Let $\mathcal{C}$ be a line arrangement in the projective plane $\mathbf{P}^2$. Define $\mathbf{P}^\sim$ and $\mathcal{C}'$ as in the example 17.6. In the search for ball quotients, in loc. cit, one first searches for weights $w(D)(D \in \mathcal{C}')$ for which the numerical conditions $\text{prop}_{orb}(\mathbf{P}^\sim) = \text{prop}_{orb}(D) = 0$ holds. They do hold for quotients of blown up balls (17.10), and similarly for quotients of blown up abelian surfaces or projective plane by finite groups.

By 17.6, vanishing of prop holds for $\mathbf{P}^\sim$, when $\mathcal{C}'$ is weighted by 1 on the pure transforms of lines in $\mathcal{C}$ and by -1 on the exceptional curves. Applying 17.4.2, one sees that if, for some system of weights on $\mathcal{C}'$ the conditions $\text{prop}_{orb}(D) = 0$ holds for all D in $\mathcal{C}'$, then $\text{prop}_{orb}(\mathbf{P}^\sim) = 0$ holds automatically.

17.13. For quotients of a ball, or of a blown up ball as in 17.10, the weights $w(D)$ (D in $\mathcal{C}'$) are reciprocal integers. It is tempting to give a geometric interpretation to the foregoing prop conditions when the integrality condition on the weights are dropped.

We assume that the divisors D in $\mathcal{C}'$ with weight $w \leq 0$ are disjoint and have negative self-intersection. Let U be the complement in $\mathbf{P}^2$ of the lines in $\mathcal{C}$.

One asks for an hermitian metric g on U such that

(17.13.1) locally, (U, g) is isometric to the complex ball.

(17.13.2) Let $D \in \mathcal{C}'$ be a divisor with weight $w > 0$.
Near a point of D which is not on any other divisor in $\mathcal{C}'$, there exists local coordinates z_1, z_2 with $z_1 = 0$ an equation for D, such that the metric is the

pull back of the standard metric on the complex ball $B \subset \mathbf{C}^2$ by (z_1^w, z_2). The pull back does not depend on the chosen branch of z_1^w.

(17.13.3) At a point x where divisors D_1, D_2 of weight $w_1, w_2 > 0$ meet, there similarly exist local coordinates (z_1, z_2) such that the metric is the pull back of the ball metric by $(z_1^{w_1}, z_2^{w_2})$.

(17.13.4) Let P be deduced from $\mathbf{P}^\sim$ by contracting to a point the divisors of negative weight, and by removing those of zero weight. The space P is the metric completion of U.

Similar definitions can be proposed to interpret "quotient of the flat space $\mathbf{C}^2$" or "quotient of the projective plane, with its Fubini metric" when weights are not reciprocal integers. In those cases, zero weights should not be allowed.

Remark 17.14. Assume that a finite automorphism group W of $(\mathbf{P}^\sim, \mathcal{C}')$ preserves the weights, and that there exists a W-invariant metric g as in 17.13. If U' is the open subset of U where W acts freely, it descends to a similar metric on U'/W, and P/W is the metric completion of U'/W. For D in $\mathcal{C}'$, let $n(D)$ be the order of the fixer of D in W. If we assume that for each D in $\mathcal{C}'$ with positive weight, $w(D)/n(D)$ is a reciprocal integer, then the orbifold P/W is a ball quotient.

If one assumes that

(a) For D in $\mathcal{C}'$ of weight w, $\frac{1}{w}\operatorname{prop}_{orb}(D) = 0$. (From this follows that $\operatorname{prop}(\mathbf{P}^\sim) = 0$).

(b) Divisors D in $\mathcal{C}'$ with weight ≤ 0 are disjoint of negative self intersection.

(c) On P the divisor $K_{orb} := K + \Sigma(1 - w(D))D$ (sum over divisors with $w(D) > 0$) is ample.

(d) For D in $\mathcal{C}'$ of positive weight, $w(D)/n(D)$ is a reciprocal integer, then methods of Kobayashi, Naruki, and Sakai should indeed prove that the orbifold P/W is a ball quotient, cf. [BHH] p. 266, [KNS].

17.15. Remark. A model for 17.13, 17.14 is provided by the hypergeometric story. Indeed, given $(\mu_i)_{1 \leq i \leq 5}$, $0 < \mu_i < 1$, $\Sigma\mu_i = 2$, hypergeometric functions provide a multivalued map w from Q into the complex ball, where Q is the complement in $\mathbf{P}^2$ of a complete quadrangle.

The pull back of the ball metric is independent of the branch of w chosen: we get a metric g on Q. It is locally a ball metric, and has on $\mathbf{P}^2$ singularities as in 17.13.

The remark 17.14 is parallel to the weakening of INT to ΣINT.

17.16. Let $W \subset PGL(\mathbf{C}^3)$ be a finite group generated by complex reflections, i.e. the image in $PGL(\mathbf{C}^3)$ of complex reflections in $GL(\mathbf{C}^3)$. We assume it irreducible, i.e. that $\mathbf{C}^3$ is an irreducible representation of the inverse image of W in $GL(\mathbf{C}^3)$. Let $\mathcal{C}$ be the set of fixed lines of complex reflections in W. We will not consider the cases leading to a Ceva arrangement since they are commensurable to those of 17.15 by [BHH] 5.6. The group W acts on $\mathbf{P}^2$, and the open set U' where it acts freely is the complement U of the lines in $\mathcal{C}$, minus a finite number of points.

Let each D in $\mathcal{C}$ be given a weight $w(D)$. We blow up $\mathbf{P}^2$ to $\mathbf{P}^\sim$, as in 17.6 and assign to the pure transform D' of D in $\mathcal{C}$ the same weight $w(D)$. If a point x is the intersection of $r \geq 3$ lines D_i in $\mathcal{C}$, we assign to the corresponding exceptional curve E the weight w such that $\mathrm{prop}_{orb} E = 0$, i.e.

$$w = \frac{1}{2}(r - 2 - \Sigma w(D_i)).$$

A case by case check ([BHH] 5.7) shows that

FACT 17.16.1. *If the system of weights w is W-invariant, for each $D \in \mathcal{C}'$ (cf. 17.3), one has $\mathrm{prop}_{orb}(D) = 0$. As a consequence (17.12), $\mathrm{prop}_{orb}(\mathbf{P}^\sim) = 0$.*

For $\mathcal{C}$ as above, let $\tau(D)$ be the number of intersection points of the arrangement on a line D of $\mathcal{C}$, let k be the total number of lines. By [BHH] p. 185, vanishing of prop in the case of constant w is equivalent to the relation

$$3\tau(D) = k + 3$$

for all lines D. This relation is easily read off the two tables of [BHH] p. 211 but we don't have a uniform proof of it.

17.17. Question 1. We use the notations of 17.16 and 17.12. If w is W-invariant and in a suitable range, does there exist on U a W-invariant hermitian metric g which is locally isometric to a complex ball, or to a Fubini, or a flat metric, and is as in 17.13 on the whole of $\mathbf{P}^\sim$? The range of w should be connected, contain the w for which conditions (a)(b)(c) of 17.14 hold (where g should be ball-like), as well as a neighborhood of $w = 1$ (here g should be Fubini like).

17.18. Such a metric g determines a "projective structure" on U: a preferred system of local charts with values in $\mathbf{P}^2$, with changes of charts given by elements of $PGL(3)$. One takes as charts composites $p \circ i$ of isometries with projective transformations.

M. Yoshida (Math. Annal. 274 (1988) 319-334) has shown the existence and uniqueness of a W-invariant projective structure on U, with an asymptotic behavior along each divisor in C' given by its weight. The weights here can be taken to be any W-invariant system of complex numbers. We note that the Ceva (cf. [BHH] p. 206) and extended Hesse configurations (cf. 15.6) can be reduced to the complete quadrangle and treated using hypergeometric functions (cf. 17.15). The remaining configurations (icosahedral, Klein, and Valentiner) are such that W acts transitively on C; hence all lines here have the same weight w.

The projective structure on U descends to one on U'/W. It gives rise to a developing map, from the universal covering of U'/W to $\mathbf{P}^2$, with a monodromy

(17.18.1) $\rho_w : \pi_1(U'/W, 0) \to PGL(3).$

For a weight w for which P/W is a ball quotient, the image of ρ_w is the corresponding lattice in $PU(1,2) \subset PGL(3)$. To determine whether it is arithmetic, it would be useful to have an a priori description of ρ_w.

For D in C', we define

(17.18.2) $q(D) = \exp(-2\pi i w(D)/n(D)).$

The monodromy around the image of D in $\tilde{\mathbf{P}}/W$ defines a conjugacy class of elements s_D in $\pi_1(U'/W, 0)$ and (17.18.2), $\rho_w(s_D)$ is conjugate to the pseudoreflection with non trivial eigenvalue $q(D)$.

If $x \in U$ has a non trivial stabilizer W_x in W, and if V is a small W_x-invariant ball around x, the fundamental group of $(V - \{x\})/W_x \subset U'/W$ is W_x. This construction defines a conjugacy class of morphisms from W_x to $\pi_1(U'/W, 0)$.

17.18.3 The restriction of ρ_w to W_x is independent of w: up to conjugacy, it is the action of W_x on the tangent space of U at x (completed to a projective space).

The group $\pi_1(U'/W, 0)$ is an extension of the group W by $\pi_1(U)$.

17.19 Question 2. Is there a unique continuous family ρ_w of conjugacy classes of representations of $\pi_1(U'/W, 0)$, depending only on the $q(D)$, satisfying (17.18.2) and (17.18.3), and which for $w = 1$ is the projection

$$\pi_1 \to W \hookrightarrow PGL(3)?$$

Uniqueness should be allowed to break down when the representation becomes reducible.

This question has an analogue for linear groups.

17.20. Let $W \subset GL(V)$ be an irreducible finite group generated by complex reflections, and let V' be the complement in V of the fixed hyperplanes of the complex reflections in W. For H the fixed hyperplane of a complex reflection in W, let s_H be the generator of the monodromy around the image of H in V/W. It is well defined up to conjugacy in $\pi := \pi_1(V'/W)$. The fundamental group π is an extension of W by the fundamental group of V', and s_H projects in W to the inverse of the generator of the fixer of H, with non trivial eigenvalue of the form $exp(2\pi i/n(H))$.

17.21. Question 3. For each conjugacy class of each hyperplane H fixed by a complex reflection, let $q_H(t)$ be a path in $\mathbf{C}^*$, starting at $exp(-2\pi i/n_H)$. Is it uniquely possible to deform with t a representation ρ_t of π, starting at $t = 0$ with the given representation of the quotient W of π, so that $\rho_t(s_H)$ is a complex reflection with non trivial eigenvalue $q_H(t)$? Uniqueness should be allowed to break down when the representation becomes reducible.

For W a Coxeter group, π is the corresponding braid group and the condition on s_H implies that the desired representation is a module over the Hecke algebra $\mathcal{H}_t$ corresponding to the $q_H(t)$. This Hecke algebra is a deformation of the group algebra of W, and a positive answer to Question 3 follows, provided that the path $q_*(t)$ avoids an hypersurface (finitely many q, if all H are conjugate).

17.22. For the icosahedral arrangement of lines in $\mathbf{P}^2$, one can deduce from this a positive answer to Question 2. This allows us to investigate the arithmeticity of the lattices in $PU(1, 2)$ corresponding to weights for which P/W is a ball quotient. The answer is that they are all arithmetic.

For the Klein configuration, one has $|W| = 168$, all the 21 lines are conjugate each with fixer of order two. Heuristic arguments, based on 17.19, lead one to expect that P/W is a ball quotient when the half weight $w/2$ takes one of the values

$$w/2 = 1/3, 1/4, 1/5, 1/6, 1/8, 1/12, 0$$

and that the corresponding lattice is arithmetic only for $w/2 = 1/3$ or 0.

Bibliography

[A]   Askey, R., *Ramanujan and hypergeometric and basic hypergeometric series*, Poona Conference.

[Au]   Auslander, L., *Bieberbach's theorem on space groups and discrete uniform subgroups of Lie groups*, I. Annals of Math. **71** (1960), 579–590; II Amer. J. Math. **83** ((1961)), 276–280.

[BPV]   Barth, W., Peters, C., Van de Ven, A., *Compact Complex Surfaces*, Springer-Verlag, 1984.

[BHH]   Barthel, G., Hirzebruch, F., Höfer, T., *Geraden-konfigurationen und Algebraische Flächen*, Vieweg, Braunschweig, 1987.

[Cor]   Corlette, K., *Archimedean superrigidity and hyperbolic geometry*, Annals of Math. **135** (1992), 165–182.

[C]   Coxeter, H.S.M., *Finite groups generated by unitary reflections*, Abh. Math. Sem. Univ. Hamburg **31** (1967), 125–135.

[DM]   Deligne, P., and Mostow, G.D., *Monodromy of hypergeometric functions and non-lattice integral monodromy*, Publ. Math. IHES No. **63** (1986), 5–89.

[EV]   Esnault, H., and Viehweg, E., *Logarithmic de Rham complexes and vanishing theorems*, Invent. Math. **86** (1980), 161–194.

[G]   Gelfand, I.M., *Collected Works, Vol. III*, Springer-Verlag, 1989.

[GG]   Gelfand, I.M., and Gelfand, S.I., *Generalized hypergeometric functions*, Dokl. Akad. Nauk SSSR **228 (2)** (1986), 279–283.

[GP]   Gromov, M. and Piatetski-Shapiro, I., *Non-arithmetic groups in Lobachevsky spaces*, Publ. IHES no. **66** (1988), 93–103.

[GS]   Gromov, M., and Schoen, R., *Harmonic maps into singular spaces and p-adic superrigidity for lattices in groups of rank one*, Publ. Math. IHES **76** (1992), 165–246.

[KNS]   Kobayashi, R., Naruki, I., and Sakai, F., *A numerical characterization of ball quotients for normal surfaces with branch loci*, Proc. Japan Acad: Ser. A, Math. Sci. **65** No. **7** (1989), 238–241.

[K]   Kodaira, K., *On compact analytic surfaces*, Collected Works, vol. III, Princeton University Press, 1975, pp. 1142–1156; Analytic Functions, Princeton University Press, 1960, pp. 121–135.

[Ku]   Kummer, E., *Uber die hypergeometrische Reihe...*, reine und angew. Math. **15** (1836), 39–83.

[La]   Lauricella, *Sur funzioni ipergeometricae a piu variabili*, Rend. di Palermo **VII** (1893), 111–158.

[Li]   Livne, R., *On certain covers of the universal elliptic curve*, Ph.D. Dissertation, Harvard (1981).

[M1]   Mostow, G.D., *Strong rigidity of locally symmetric spaces*, Ann. of Math. Studies, Princeton Univ. Press **78** (1973), 195.

[M2]   ――――, *A remarkable class of polyhedra in complex hyperbolic-space*, Pac. J. Math. **86** (1980), 171–276.

[M3]   ――――, *Generalized Picard lattices arising from half-integral conditions*, Publ. Math. IHES No. **63** (1986), 91–106.

[M4]   ――――, *On discontinuous action of monodromy groups on the complex n-ball*, J. of AMS **1** (1988), 555–586.

[Pf]   Pfaff, J.F., *Observationes analyticae ad L. Euler institutiones calculi integrali*, Nova Acta Acad. Sci. Petropolitanae **11** (1797), 38–57.

[Pi]   Picard, E., *Sur une extension aux fonctions de deux variables du problème relatif aux fonctions hypergéométriques*, Ann. E.N.S. **10** (1881), 305–321.

[S]   Sauter, J.K., *Isomorphisms among monodromy groups and applications to lattices in PU*(1,2), Yale U. Dissertation (1988): Pacific J. Math. (1990).

[Sc]   Schwartz, L., *Théorie des Distributions, t 1*, Hermann, Paris, 1957.

[ST]   Shepard, G.G., Todd, I.A., *Finite unitary reflection groups*, Canadian J. Math. **6** (1954), 274–304.

[T]   Terada, T., *Problème de Riemann et fonctions automorphes provenant des fonctions hypergéométriques de plusieurs variables*, J. of Math. of Kyoto Univ. **13-3** (1973), 557–578.

[Y]   Yoshida, M., *Orbifold-uniformizing differential equations II, arrangements defined by 3-dimensional primitive unitary reflection groups*, Math. Anal. **274** (1988), 319–334.

EGA III (Seconde Partie), Eléments de Géométrie Algébrique par A. Grothendieck et J. Dieudonné. Etude Cohomologique des faisceaux cohérents, Publ. Math. IHES, No. 17, (1963).

SGA 4, Séminaire de Géométrie Algébrique du Bois Marie 1963-64, dirigé par M. Artin, A. Grothendieck, J.L. Verdier, Lecture Notes in Math. 269, 270, 305. Springer Verlag 1973, Exposé XI in LN 305, pp. 64-78.

SGA $4\frac{1}{2}$ Séminaire de Géométrie Algébrique du Bois Marie, Cohomologie étale, Lecture Notes in Math. No. 569, Springer-Verlag (1977).